3

TESI

tesi di perfezionamento in Matematica per la tecnologia e l'industria sostenuta il 18 Dicembre 2006

Edoardo Sinibaldi
Scuola Superiore Sant'Anna
CRIM Lab

Implicit preconditioned numerical schemes for the simulation of three-dimensional barotropic flows

Edoardo Sinibaldi

Implicit preconditioned numerical schemes for the simulation of three-dimensional barotropic flows

EDIZIONI
DELLA
NORMALE

ISBN: 978-88-7642-310-9

*...to **Roberta**,*

who patiently supported me in this work.
With never-ending Love.

Contents

Introduction

Motivation of the study

The present thesis documents the starting efforts made for constructing a numerical frame aimed at simulating propellant flows occurring in the feed turbo-pumps of modern liquid propellant rocket engines. More precisely, the numerical simulation of the three-dimensional (hereafter 3D as well) unsteady propellant flows around the axial inducers which are part of the aforementioned turbo-machines (see Section 1.1) is the long-term goal of the present research.

The suction performance and, consequently, the global performance of the rocket engine significantly depend on the flow pattern within the turbo-pumps [9]; hence, it is of interest to understand and control their hydrodynamics, in order to conceive a correct design. Besides the experimental investigations, undoubtedly expensive as well as quite dangerous when dealing with many liquid propellants [97], the numerical simulation may provide a deep insight into the flows under consideration at a generally affordable cost, thus motivating the development of suitable numerical tools. Nevertheless, several factors make it challenging to accurately and efficiently simulate the considered liquid flows. Firstly, the severe weight and size constraints to which the considered engines are subjected impose high rotor speeds which, in turn, systematically entail the occurrence of cavitation phenomena (see Section 1.2). The simultaneous presence of the pure liquid (which is almost incompressible) and the liquid-vapour mixture (which behaves like a highly compressible fluid) dramatically changes the local flow properties at the unknown interface, thus rendering common numerical methods hardly applicable. Furthermore, the very complex geometry of the axial inducers adds to the difficulty of the problem. Indeed, it imposes to adopt huge computational grids which, if unstructured, generally do no permit to straightforwardly extend some well-known numerical techniques and, in any case, require efficient, hopefully parallel, algorithms to be defined.

In view of the aforementioned considerations, the subject of the present study seems to possess aspects of interest from both an academic and an industrial point of view. As far as the latter point is concerned, a significant part of the documented research activity has been funded by the Italian Space Agency (ASI) under a 16 month industrial program, namely the FAST2 (Future Advanced Space Transportation Technologies) project.[1]

Overview and choice of the numerical methodology

In consideration of the fact that, under typical operational conditions, cavitation phenomena can take place within the aforementioned flows, it is necessary to select a suitable cavitation model as a first step in the development of the numerical tool under consideration. Indeed, the cavitation model specifically adopted directly affects the mathematical formulation of the problem (through the closure of the governing equations) and therefore its numerical discretization. A concise overview of the current cavitation models is reported in Section 1.3. For the present purposes it suffices to mention that almost all the formulations used for computations of industrial interest are based on *equivalent fluid models*, namely:

(BH) barotropic *homogeneous flow models* (see Section 1.3), according to which the pressure and the density are linked to each other by an invertible relation within both the pure liquid and the cavitating mixture. Examples of this approach may be found, for instance, in [16, 21, 22, 23, 24, 25, 45, 54, 76, 78, 82, 96, 104] and [105];

(DS) "dual species" models (a particular class of *non-homogeneous flow models*, see Section 1.3) in which a convection equation for the volume or mass void fraction is introduced, explicitly accounting for the mass transfer at phase transition. Examples of this approach may be found, for instance, in [3, 46, 47, 58, 70, 86, 87, 89] and [116].

With the exception of [46] and [47], the aforementioned works deal with structured grids (possibly involving generalized curvilinear coordinates as for [58] and [89]) and adopt a finite volume spatial discretization (see Sections 3.1.1 and 5.1). Some of them, belonging either to the (BH) or to the (DS) class, implement a pressure-based approach typically extending well-known pressure-correction algorithms originally conceived for incompressible flows (*e.g.* the SIMPLE algorithm [74] or some related

[1] The author joined the project as a collaborator of the Aerospace Engineering Dept. of the University of Pisa which, in turn, was involved in the program as subcontractor of the Italian Aerospace Research Center (CIRA).

variants, like PISO) in order to suitably cope with the compressible, cavitating flow sub-domains. Other works, instead, adopt a density-based approach which modifies (by preconditioning techniques, see *e.g.* Section 3.4) common algorithms originally conceived for compressible flows so as to account for the very weak liquid compressibility. In most cases, the convective component of the numerical flux function (see the relevant paragraph in Section 3.1.1) is discretized by upwinding: typically a TVD scheme (see *e.g.* [98]) or, for the (BH) class, an artificial dissipation approach (see *e.g.* [53]) is exploited. Conversely, the diffusive component of the numerical flux (if any) is discretized by central differencing for almost all the considered works. As far as the time-advancing is concerned, both explicit and implicit techniques (see Section 3.1.2) are considered but only the latter allow for the construction of efficient schemes. A dual time-stepping is adopted in certain works (*e.g.* [22, 23, 24] and [58]), in which a suitable preconditioning technique as well as an under-relaxation of the density are introduced in order to speed-up the convergence of the internal iterations. It is worth noticing that only Coutier-Delgosha and coworkers currently manage to compute cavitating flows in inducers with a certain degree of accuracy[2] while other researchers only succeed in dealing with less ambitious (even if still challenging) applications like nozzle and hydrofoil flows.

On the basis of the literature reviewed at the beginning of the research project here documented, both the pressure-based and the density-based approach seemed to possess points of strength as well as weaknesses, so that there was no clear advantage in *a priori* preferring one to the other. Because of this point, a resource-driven choice was made. In particular, a density-based approach was selected, suitable for incorporation within a numerical framework for the simulation of compressible flows, which was available to the research group.[3] The numerical tool under consideration is the AERO code (see *e.g.* [32, 35] and [71]), derived from a collaboration between the French national institute for research in computer science and control (INRIA, "Institut National de Recherche en Informatique et en Automatique") and the University of Boulder (Colorado, USA).

The AERO code discretizes both the laminar and the turbulent Navier-Stokes equations (written in conservation form) for ideal gases; in the latter case either a Reynolds-averaged formulation (closed by several turbulence models) or a LES (Large-Eddy Simulation) formulation can be

---

[2] The most advanced results, reported in *e.g.* [22] and [24], were not published at the time the research project here documented started.

[3] Namely, the CFD group of the Aerospace Engineering Dept. of the University of Pisa.

adopted. Moreover, it permits to simulate one-way fluid-structure inter-
actions. Space and time discretizations are kept separate ("method of
lines"). The space discretization is carried out by a mixed finite volume-
finite element formulation (see *e.g.* [85]) based on tetrahedral unstructured grids. The first-order approximation of the convective fluxes is obtained by means of the Roe flux function [84]; higher order extensions
are based on a MUSCL-like reconstruction [108] conceived for unstructured grids (see *e.g.* [28]). As for the diffusive fluxes, P1 finite elements
are exploited. A Roe-Turkel preconditioning technique is adopted for
low Mach number (*i.e.* nearly incompressible) flows, which can be exploited for unsteady simulations as well (see *e.g.* [42] and [112]). As far
as the time discretization is concerned, either explicit or implicit time-
advancing strategies are available. In the former case, a low-storage 4−th
order Runge-Kutta scheme is adopted while, in the latter one, a linearized
implicit scheme designed for the ideal gas state law (see *e.g.* [36]) is implemented and the extension to the second order in time is achieved by
a "defect-correction" strategy [67]. An efficient implementation of the
whole numerical framework is achieved by means of a message-passing
(MPI-1 standard) parallelization strategy.

In order to construct a numerical solver for cavitating flows in complex
geometries starting from AERO, only the inviscid portion of the laminar
governing equations (*i.e.* the Euler equations) has been considered (see
Section 2.2 for the rationale). A barotropic homogeneous flow cavitation
model (able to take into account thermal cavitation effects and, possibly,
the concentration of the active cavitation nuclei) as well as a barotropic
state law for the pure liquid have been chosen, thus providing a unified
barotropic state law for the working fluid (see Section 4.1).

Note 1. For the sake of generality, only a very few, physically-based,
constraints have been imposed on the considered barotropic state law (see
Section 1.5). Hence, all the proposed numerical ingredients (with the
only exception of the Godunov numerical flux discussed in Section 3.2,
which additionally requires the state law to be convex) can be applied to
generic barotropic laws.

The adopted barotropic formulation permits to decouple the energy
balance from the rest of the governing equations and therefore a "reduced" system only comprising the mass and the momentum balance has
been considered. The chosen state law directly affects the space discretization of the AERO solver through the definition of the Roe numerical flux. Moreover, it also affects the time discretization, since the linearized implicit time-advancing strategy originally appearing in AERO
directly exploits the first-order homogeneity of the analytical flux, which

holds for the ideal gas case but not for the barotropic one (see Section 3.5). In addition, the adopted state law indirectly appears within the Roe-Turkel preconditioning strategy as well (see Section 3.4.3). In view of the above considerations, a Roe numerical flux function suitable for generic barotropic flows has been proposed as a basic numerical ingredient. The spatial accuracy of the numerical solution obtained by applying the resulting finite volume scheme to nearly-incompressible flows has then been addressed, by performing the asymptotic analysis originally proposed in [42]. It has been shown that the preconditioning technique proposed in [42] for the ideal gas case can be extended to the barotropic one. Carried out one-dimensional (hereafter 1D as well) numerical experiments have confirmed the predicted accuracy problems occurring at low Mach numbers as well as the effectiveness of the considered preconditioning strategy [91]. However, it has been also observed that the preconditioning at hand restricts the stability region of common explicit time-advancing schemes (see Section 3.4.4), thus decreasing their efficiency [91]. In order to counteract this problem, a linearized implicit time-advancing strategy has been proposed, which only relies on the algebraic properties of the Roe flux function (and therefore it is well suited to a variety of problems) and which can be applied to the preconditioned formulation as well. As shown by the aforementioned 1D numerical experiments, the implicit scheme allows for a very efficient time-advancing to be performed when non-cavitating flows are considered; however, when cavitation occurs, the efficiency of the implicit scheme is noticeably reduced. A 3D numerical method has been subsequently derived from the considered 1D techniques by exploiting the tensorial character of the governing equations. In view of the time-schedule imposed by the supporting industrial program, the proposed 3D numerical method has been directly implemented within the AERO mainframe and the resulting numerical tool has been validated by considering the water flow around a NACA0015 hydrofoil. The proposed 3D numerical method has finally been extended so as to deal with rotating frames and the corresponding implementation has been validated by considering the water flow around an axial turbo-pump inducer. The efficiency issues originally noticed in a 1D context have been systematically observed in the 3D case as well. A more systematic investigation of the aforementioned 1D numerical ingredients has been consequently started. In this context, the exact solution of a 1D Riemann problem involving a generic convex barotropic state law has been constructed (based on classical elements of the theory of hyperbolic partial differential equations), to be exploited for defining exact benchmarks for the analysis and the validation of the considered 1D numerical schemes, also when considering cavitating test-cases. A

Godunov numerical flux function based on the aforementioned exact solution has been defined as well.

Thesis outline

- In Chapter 1 the considered industrial problem is presented. More
  in detail, once chosen a suitable (barotropic homogeneous flow) cavitation model, a concise statement of the industrial problem under
  consideration is reported, highlighting the generality of the chosen
  barotropic state law;
- in Chapter 2 a hierarchy of governing equation is presented, each of
  which is considered at a subsequent stage of the discussion. Once
  underlined the hyperbolic character of the systems at hand, the attention is focused on the 1D Riemann problem (hereafter RP as well); a
  constructive procedure for determining its exact solution when considering a generic convex barotropic state law is proposed;
- in Chapter 3 all the proposed 1D numerical ingredients are presented.
  After introducing some basic material on the numerical discretization,
  a Godunov numerical flux for generic convex barotropic state laws
  as well as a Roe numerical flux for generic barotropic state laws are
  proposed. The behaviour of the considered semi-discrete formulation
  (based on a finite volume approach involving the Roe numerical flux)
  dealing with nearly-incompressible flows is addressed and a suitable
  preconditioning strategy is presented following [42]. Moreover, a linearized implicit time-advancing strategy is proposed, only relying on
  the algebraic properties of the Roe numerical flux function and therefore applicable to a variety of problems. Finally, the linearized implicit strategy is extended so as to deal with the preconditioned numerical flux function. All the proposed ingredients are validated against
  exact (namely, solutions to 1D RPs) or nearly-exact benchmarks. The
  issue of the efficiency of the considered scheme when dealing with
  discontinuous flow fields (mimicking cavitating conditions) is put forward;
- in Chapter 4 the barotropic state law specifically adopted for the subsequent simulation of the industrial test-cases is introduced. Moreover, an illustrative 1D numerical experiment involving cavitation
  phenomena is considered, in order to highlight some difficulties that
  are systematically encountered when dealing with the chosen cavitation model (or similar ones);
- in Chapter 5 the proposed (preconditioned) Roe numerical flux is extended to the 3D case. Moreover, the discretization of the domain
  as well as the numerical treatment of the convective fluxes are dis-

cussed. The considered 3D numerical method is then extended to rotating frames. Finally, the linearized implicit time-advancing originally proposed in a 1D context is extended to the 3D rotating case;

- in Chapter 6 the applications of the proposed 3D numerical method, namely the simulation of water flows around a hydrofoil and an axial turbo-pump inducer, are presented. In the former case, a quantitative appraisal is given for both non-cavitating and cavitating conditions, based on available experimental data. In the latter one, a qualitative appraisal is given for a non-cavitating flow;

- in Chapter 7 the main achievements of the present study, as well as its open questions, are summarized, together with some research perspectives.

Auxiliary material (*e.g.* some mathematical derivations and proofs) is finally reported in the Appendices A and B, for ease of presentation.

Note 2. No details are given in the present document concerning the implementation of the proposed numerical schemes within the aforementioned parallel numerical frame, because they are behind the scope of the discussion.

Related scientific documentation

Some numerical ingredients discussed in Sections 3.3, 3.4, 3.5 and Chapter 5, as well as the applications reported in Chapter 6 have been documented through:

- the international publications [6] and [93];
- the INRIA research report [91];
- the proceedings of the international conferences [92] and [94];
- the proceedings of the national conference [90].

Other issues (*e.g.* those presented in Sections 2.5 and 3.2) originally appear in the present document.

Chapter 1
Industrial problem

The present study is aimed at developing a numerical method suitable for the numerical simulation of propellant flows occurring in the feed turbo-pumps of modern liquid propellant rocket engines. More in detail, the numerical simulation of 3D unsteady liquid flows through axial inducers is the long-term goal of the present research.

In Sections 1.1 to 1.4 several issues are concisely presented, concerning the physical modelling of the industrial problem of interest. Among the wide variety of technical and conceptual aspects potentially arising during the discussion, only those required by the subsequent treatment are introduced, for ease of presentation. In Section 1.5 the general form of the state law adopted for the working fluid is defined. Finally, in Section 1.6, a statement of the industrial problem under consideration is reported, based on the material presented through the previous sections.

1.1. Axial inducers

Axial flow inducers are hydraulic devices suitably designed to improve the performance of the (usually centrifugal) pumps they are attached to, by increasing the inlet pressure to the pump to a level at which it can operate without excessive loss of performance due to cavitation (see Section 1.2). Typically they consist of an axial flow stage, like that one shown in Figures 1.1 and 1.2, placed just upstream of the inlet to the main impeller. They are designed to operate at small incidence angles and to have thin blades so that the perturbation to the flow is small in order to minimize the production of cavitation and its deleterious effect upon the flow: the objective is to raise the pressure very gradually to the desired level [1] [9].

---

[1] The reason why the design incidence angle is not zero is that, under these conditions, cavitation could form on either the pressure or the suction surfaces of the blades or it could oscillate between the two. It is preferable to use a few degrees of incidence to eliminate this uncertainty and ensure suction surface cavitation [9].

Axial inducers can be "shrouded" or "unshrouded". In the former case, there is no gap between the tip of the blades and the external case while in the latter one such a gap is present. An example of an unshrouded axial inducer is shown in Figures 1.1 and 1.2. The shrouded geometry makes the inducer more robust with respect to cavitation instabilities.[2] In addition, the absence of the gap prevents the creation of very complex secondary flows (synthetically referred to as "tip leakage") which affect the inducer fluid dymanics [59], generally weakening the inducer pumping performance. In spite of their attractive features, the manufacturing process required by shrouded inducers is very complex (and expensive); hence, most inducers are nowadays unshrouded [9].

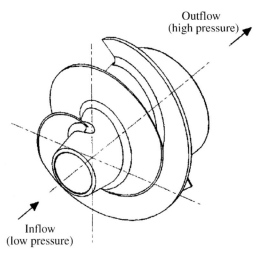

Outflow
(high pressure)

Inflow
(low pressure)

Figure 1.1. Schematic of a two-bladed unshrouded helical inducer. The far-field inflow is aligned with the rotation axis; the (swirled) outflow is directed towards the main pump impeller (not shown).

As for the vast majority of modern turbo-machines typical of space propulsion applications, also for axial inducers the very strict weight and size constraints impose, for a given power, a high rotational speed. This, in turn, entails high tip speeds and paves the way for cavitation to take place. A certain understanding of cavitation phenomena as well as suitable modelling techniques are therefore needed in order to describe the liquid flows within this kind of machines, even when they are not expressly designed to operate in cavitating conditions.

---

[2] For an exhaustive treatment of cavitation instabilities in inducers the interested reader can refer to the work of Y. Tsujimoto, not reported in the bibliography because beyond the scope of the present work.

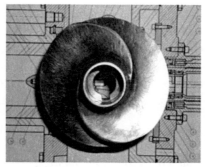

Figure 1.2. Side and front views (left to right) of the two-bladed unshrouded helical inducer sketched in Figure 1.1.

1.2. Cavitation

Cavitation is a complex fluid dynamic phenomenon, involving the extremely rapid growth and subsequent collapse of liquid cavities originating from weak spots (cavitation nuclei) when the pressure falls below the saturation value for a sufficiently long time for the nuclei to become active [10]. The relative abundance and susceptibility of nuclei in the bulk of the liquid and on its low pressure boundaries determines the dispersed or attached form of cavitation.

A major difficulty in the analysis of cavitating flows is the presence of free surfaces, whose shape, location and evolution are not known *a priori* and must in principle be obtained as part of the solution of the flow field. Cavitating flows are therefore intrinsically unsteady on a length scale comparable to the cavity size and often also on the global (macroscopic) scale, especially in internal reverberating flows. The dynamic nature of cavitation, with the occurrence of appreciable inertial effects in the liquid and rate-controlled evaporation/condensation at the interface, adds to the complexity of the phenomenon, since thermodynamic equilibrium is not satisfied and the usual barotropic behaviour of common fluids[3] should in principle be replaced by a differential relation between the local density and pressure. Therefore cavitation poses formidable obstacles in terms of both physical and numerical modelling.

1.3. Cavitation modelling

Current models for the description of cavitating flows can be classified as follows:

---

[3] *I.e.* the possibility of expressing the thermal state law of a certain fluid by means of a one-to-one correspondence between density and pressure.

- *free streamline models*, where the cavity region is separated by a sharp interface from the region occupied by the pure liquid (*e.g.* [117]);
- *equivalent fluid models*, where volume, time or ensemble averaging is used to account for the presence of two phases (*e.g.* [50]);
- *direct simulation models*, where the coupled Navier-Stokes equations of the two phases are solved simultaneously.

In turn, equivalent fluid models can be divided in:

- *homogeneous flow models*, where the macroscopic features of cavitation are represented in terms of a single-phase fluid whose properties are obtained by introducing suitable simplifying assumptions (*e.g.* [30]);
- *non-homogeneous flow models*, based on the separate characterization of the two phases with the relevant interaction terms (*e.g.* [2] and [70]);
- *non-homogeneous flow models with cavity dynamics*, similar to the previous ones except for the inclusion of the evolutionary effects connected to the transfer of mass, momentum and energy between the two phases (*e.g.* [15], [57] and [77]).

None of these models is free from inherent limitations. Free streamline models, where a well-defined interface separates the pure liquid from the cavity region occupied by the non-condensed phase, introduce prohibitive complications in 3D configurations and are not realistic in the thermal cavitation conditions typical of cryogenic propellants of rocket motors, where travelling bubble cavitation prevails [9]. On the other hand, direct simulation methods are extremely demanding in terms of computational resources and their superior accuracy is eluded in practice by the uncertain knowledge of the initial state of the system, especially the nature, concentration and susceptibility of cavitation nuclei.

This brief overview indicates that the successful choice of a model for simulating cavitation in technical applications must be based on careful consideration of the final objectives and implementation constraints, in order to exploit all opportunities to simplify the formulation of the problem by including only the essential physical phenomena.

1.4. Choice of the cavitation model

The typical requirements of space propulsion applications for the analysis of propellant feed turbo-pumps put especial prize on the suction and dynamic performance of the machine rather than on its resistance to erosion and other long-term effects of cavitation, which are typically a major concern in other applications. Fortunately these properties are essentially related to the large-scale characteristics of the flow field. The local behaviour of the cavities, on the other hand, mainly controls aspects such as erosion, high frequency vibrations and noise, which are less important in liquid propellant rocket engines in view of their limited expected life time.

These considerations indicate the opportunity of choosing an equivalent fluid model where the fine details of the cavity growth and collapse are neglected, and the cavitating flow is visualized in terms of a single fluid, whose properties are obtained by introducing suitable simplifying assumptions.

In cavitating liquids with relatively high vapour pressures (like most cryogenic propellants) the transfer of heat at the interface represents the most important interaction between the two phases because of its influence on the cavity pressure ("thermodynamic effect") and indirectly on the other flow variables. Conversely, in these flows mass and momentum exchanges usually play a comparatively minor role. In view of these considerations, pressure and velocity differences between the two phases can safely be neglected and the choice of an homogeneous flow model explicitly accounting, at least approximately, for thermal cavitation effects seems to be an efficient approach to the simulation of cavitating flows for performance predictions in space propulsion applications [26].

Among the cavitation flow models meeting the above requirements, that one recently proposed by d'Agostino and coworkers [27] deserves special mention. In this model the liquid/vapour mixture behaves isentropically, so that it is possible to use the mixture energy balance in order to evaluate the mass interaction term accounting for evaporation/condensation phenomena between the two phases and ultimately derive a monotonic constitutive relation between the density ρ and the pressure p of the cavitating mixture (*i.e.* a barotropic thermal state law). In addition, the model naturally accounts for the effects of thermal cavitation by exploiting the specific properties of the thermally-controlled dynamics of cavitating spherical bubbles. Finally, if required, the effects of the active nuclei concentration in the liquid phase can be readily incorporated in the model. Only the essential features of the model are reported below, further details being given in Section 4.1.2.

In order to keep some degree of generality, the state law provided by the chosen cavitation model is expressed as a generic curve of the form:

$$p = p_{\text{cav}}(\rho) \quad , \quad \rho \in [\rho_{\text{min}}, \rho_{\text{Lsat}}]. \tag{1.1}$$

The lower bound of the domain, ρ_{min}, is constrained by some simplifying assumptions on which the cavitation model is based (see Section 4.1.2) while the upper bound, ρ_{Lsat}, is the liquid saturation density at a given temperature[4] and represents the transition to the pure liquid regime. The physical foundations of the model ensure the strict positivity of ρ and p, which can therefore be interpreted as the "usual" density and pressure of an equivalent fluid in the sense of classical fluid mechanics. Furthermore, the strict positivity of the derivative $dp/d\rho$ is also guaranteed, thus satisfying a classical thermodynamic stability requirement [12]

By virtue of the isentropic approximation adopted while deriving the cavitation model, it is possible to interpret the following entity:

$$a := +\sqrt{\frac{dp}{d\rho}} \qquad (1.2)$$

as a mixture "sound speed", in analogy with classical fluid mechanics. More in general, (1.2) is adopted for defining the sound speed of a generic barotropic fluid through the present document.

A common practice is to juxtapose the barotropic state law provided by a homogeneous cavitation model with another one describing the pure liquid,[5] in order to obtain a unified model for the working fluid,[6] which may or may not cavitate depending on the flow conditions. This approach seems to represent a good compromise between computational cost and accuracy, and it is adopted for the numerical simulation of the industrial test-cases reported in Chapters 4 and 6.

Despite their formal simplicity, however, considerable difficulties are still encountered in implementing physically-based unified barotropic models in a workable simulation tool for the prediction of cavitating flows. Indeed, the local presence of two phases dramatically reduces the sound speed of the mixture:[7] both nearly-incompressible zones (pure liquid) and regions where the flow may easily become highly supersonic (liquid-vapour mixtures) are present in the flow and need to be solved simultaneously. The sound speed variation at cavitation inception, in particular, is exceedingly abrupt (see Section 4.1.2); it originates discontinuities within the flow field that, together with the strong shocks occurring

---

[4] The liquid is supposed to be at constant temperature; hence, ρ_{Lsat} is properly defined.

[5] Usually, the starting point is a given model for the pure liquid to be coupled with a consistent homogeneous flow cavitation model for the mixture. This perspective has been somehow twisted in the present discussion for ease of presentation.

[6] The chosen cavitation model, in particular, allows a smooth junction (i.e. continuity of p and a, see Section 4.1.2) to be defined at the transition point ρ_{Lsat}.

[7] For instance, in a water-vapour mixture at 20°C, $a \approx O(10^3)$ m/s towards the pure liquid limit, it abruptly decreases to $O(10^{-1} \div 10^0)$ m/s in the mixture before rising back to $O(10^2)$ m/s towards the pure vapour limit.

in the recondensation at the cavity closure, add to the complexity of the phenomenon. It is therefore evident that specifically designed numerical schemes must be introduced in order to handle this situation.

1.5. Definition of the state law

Let $D_\rho := [\rho_{min}, \rho_{sup})$ denote a density domain. In view of the considerations reported in Section 1.4, a generic state law is assumed for the working fluid, of the form:

$$p = p(\rho) \quad , \quad \rho \in D_\rho \tag{1.3}$$

with:

$$\frac{dp}{d\rho}(\rho) > 0 \quad , \quad \rho \in D_\rho . \tag{1.4}$$

No loss of generality is introduced by choosing ρ as the independent variable (p would be equivalently acceptable). According to (1.4) the pressure is allowed to vary within the domain $D_p := [p_{min}, p_{sup})$, with $p_{min} = p(\rho_{min})$ and $p_{sup} = p(\rho_{sup})$. Clearly, for the state law to be meaningful from the viewpoint of classical fluid mechanics, the following conditions must be verified as well:

$$\rho_{min} > 0 \quad , \quad p_{min} > 0 . \tag{1.5}$$

1.6. Statement of the industrial problem

In consideration of the material introduced up to this point, it is possible to state that the present study is aimed at "*developing a numerical method for simulating the flow of a fluid showing the constitutive behaviour defined in Section 1.5 around 3D geometries of the type of those shown in Section 1.1*".

Clearly, it is of primary interest to simulate non-cavitating flows at a first stage, and then to be able to cope with the additional difficulties introduced by cavitation phenomena. In this spirit, the above statement keeps a certain degree of generality: the possibility of simulating a pure liquid or a cavitating mixture (by a homogeneous cavitation model) is completely delegated to the specific state law and, of course, to the actual flow conditions.

Chapter 2
Mathematical formulation

A natural framework for the mathematical formulation of the industrial problem introduced in Chapter 1 is that one of classical fluid mechanics. Within this scope, several sets of governing equations are introduced in Section 2.2, to be closed by the barotropic state law discussed in Section 1.5. Each of them is representative of a certain type of approximation and it is exploited at a specific point during the subsequent development of numerical methods. Once recalled some basic issues related to hyperbolic systems and integral solutions in Section 2.3, some attention is devoted to the Riemann problem in Section 2.4 because of the key role its solution plays in the set up of modern numerical methods for fluid dynamics. Finally, in Section 2.5 the ingredients presented in Section 2.4 are exploited to solve the Riemann problem associated with a generic convex barotropic state law.

General conventions concerning the notation adopted throughout the present document, as well as some relevant definitions, are firstly introduced in Section 2.1.

2.1. Notation and preliminary definitions

(C1) A bold lowercase symbol like $\mathbf{v}$ denotes a matrix in $\mathbb{R}^{n \times 1}$, hereafter referred to as a vector in $\mathbb{R}^n$. A bold uppercase symbol like $\mathbf{M}$ denotes a matrix in $\mathbb{R}^{n \times n}$. For the present purposes $n \in \{2, 3, 4\}$. In particular, $\mathbf{0}$ denotes the null vector while $\mathbf{I}$ and $\mathbf{O}$ respectively denote the identity matrix and the null matrix.

(C2) Let a, b, c,... o and p be real numbers. Moreover, let $\mathbf{v}_1 \in \mathbb{R}^3$, $\mathbf{v}_2 \in \mathbb{R}^3$, $\mathbf{M}_1 \in \mathbb{R}^{3 \times 3}$, $\mathbf{v}_3 \in \mathbb{R}^4$ and $\mathbf{M}_2 \in \mathbb{R}^{4 \times 4}$ admit the following representation:

$$\mathbf{v}_1 = \begin{pmatrix} b \\ c \\ d \end{pmatrix} \quad , \quad \mathbf{v}_2 = \begin{pmatrix} e \\ i \\ m \end{pmatrix} \quad , \quad \mathbf{M}_1 = \begin{pmatrix} f & j & n \\ g & k & o \\ h & l & p \end{pmatrix}$$

$$\mathbf{v}_3 = \begin{pmatrix} a \\ b \\ c \\ d \end{pmatrix} \quad , \quad \mathbf{M}_2 = \begin{pmatrix} a & e & i & m \\ b & f & j & n \\ c & g & k & o \\ d & h & l & p \end{pmatrix}.$$

Then, the following compact notation is understood:

$$\mathbf{v}_3 = \begin{pmatrix} a \\ \mathbf{v}_1 \end{pmatrix} \quad , \quad \mathbf{M}_2 = \begin{pmatrix} a & \mathbf{v}_2^T \\ \mathbf{v}_1 & \mathbf{M}_1 \end{pmatrix}$$

where $\mathbf{v}_2^T$ denotes the transpose of $\mathbf{v}_2$.

(C3) The "$\cdot$" (centred dot) symbol indicates the common matrix-matrix multiplication.

(C4) Let $\mathbf{M} \in \mathbb{R}^{n \times n}$ be diagonalizable with real eigenvalues λ_h, $h = 1, \ldots, n$. Then:

$$\mathbf{M} = \mathbf{T} \cdot \Lambda \cdot \mathbf{T}^{-1}$$

where Λ is diagonal:

$$\Lambda := \mathrm{Diag}\,(\lambda_1, \ldots, \lambda_n)$$

and $\mathbf{T}$ is a matrix whose columns are given by the (right) eigenvectors of $\mathbf{M}$. Once defined the application of the usual absolute value $|\cdot|$ to Λ as follows:

$$|\Lambda| := \mathrm{Diag}\,(|\lambda_1|, \ldots, |\lambda_n|)$$

it is possible to introduce the following definitions, extensively used in the sequel:

$$|\mathbf{M}| := \mathbf{T} \cdot |\Lambda| \cdot \mathbf{T}^{-1} \tag{2.1}$$

$$\mathbf{M}^\pm := \frac{1}{2} \left(\mathbf{M} \pm |\mathbf{M}| \right). \tag{2.2}$$

(C5) The symbol $\|\mathbf{v}\|$ denotes the L_2 norm of the vector $\mathbf{v}$. Unit vectors (henceforth called versors as well) are marked by a top hat, *e.g.* $\hat{\mathbf{v}}$.

(C6) Symbol $t \in [0, \infty)$ always denotes time.

(C7) The partial derivative of the scalar f with respect to the scalar v is denoted by $\partial_v f$. The partial derivative of the vector $\mathbf{f} \in \mathbb{R}^n$ with respect to the vector $\mathbf{v} \in \mathbb{R}^n$ is denoted by $\partial_{\mathbf{v}} \mathbf{f}$; it is the usual Jacobian matrix in $\mathbb{R}^{n \times n}$, whose ij-th component is given by $\partial_{v_j} f_i$, where f_i and v_j here denote the i-th component of $\mathbf{f}$ and the j-th component of $\mathbf{v}$, respectively. Consistently, the derivative $\partial_v \mathbf{f}$ is manipulated as a matrix in $\mathbb{R}^{n \times 1}$ (*i.e.* a vector in $\mathbb{R}^n$) while the derivative $\partial_{\mathbf{v}} f$ is manipulated as a matrix in $\mathbb{R}^{1 \times n}$ (*i.e.* the transpose of a vector in $\mathbb{R}^n$).

2.2. Governing equations

The Euler equations of classical fluid mechanics, which describe the flow of a compressible and inviscid fluid [88], are chosen as governing equations. The inviscid approximation seems to be justified, at least at a first stage, by the fact that [26]:

- viscous stresses are usually negligible with respect to the huge dynamic actions typical of modern hydraulic turbo-machinery for space propulsion systems;
- in these applications, viscous dissipation plays a minor role in the energy balance, if compared to the contribution due to heat conduction.

More in detail, the Euler equations for a force-free flow are considered, since also the body forces are usually negligible with respect to the dynamic actions under consideration. Furthermore, by virtue of the barotropic state law (1.3), the energy balance is decoupled from the others (*i.e.* mass and momentum) [88]; hence, a "reduced" set of equations is considered.

Once introduced the main system of governing equations in both inertial and rotating frames in Sections 2.2.1 and 2.2.2, respectively, several simplified systems are concisely reported in Sections 2.2.3 to 2.2.5. The hierarchical structure of the presented systems is finally discussed in Section 2.2.6, with the aim of highlighting the degree of approximation of each of them. Neither boundary nor initial conditions are considered at this stage of the discussion.

2.2.1. 3D equations

Let $\hat{\mathbf{e}}^{(k)}$, $k = 1, 2, 3$ be the k-th versor of a chosen Cartesian orthogonal frame associated with the physical (Euclidean) space. Moreover, let $\mathbf{u} \in \mathbb{R}^3$ indicate the flow velocity, with k-th component u_k. Finally, let $\mathcal{V}$ be an arbitrary (regular) space domain having (regular) boundary $\mathcal{S}$ with unit outer normal $\hat{\mathbf{n}}$. Then, the conservation of mass and momentum [88] within $\mathcal{V}$ can be expressed as follows:

$$\partial_t \int_{\mathcal{V}} \mathbf{q} \, dV + \int_{\mathcal{S}} \left(\sum_{k=1}^{3} \hat{n}_k \, \mathbf{f}^{(k)} \right) dS = \mathbf{0} \qquad (2.3)$$

where $\hat{n}_k$ represents the k-th component of $\hat{\mathbf{n}}$, the vectors $\mathbf{q}$ and $\mathbf{f}^{(k)}$ are defined as follows:

$$\mathbf{q} := \begin{pmatrix} \rho \\ \rho \mathbf{u} \end{pmatrix} \qquad (2.4)$$

$$\mathbf{f}^{(k)} := u_k \, \mathbf{q} + p \begin{pmatrix} 0 \\ \hat{\mathbf{e}}^{(k)} \end{pmatrix} \tag{2.5}$$

and the pressure p is related to the density ρ by means of the barotropic law (1.3). Regular solutions of (2.3) also satisfy its corresponding differential form, namely:

$$\partial_t \, \mathbf{q} + \sum_{k=1}^{3} \partial_{x_k} \mathbf{f}^{(k)} = \mathbf{0} \tag{2.6}$$

where x_k denotes the k-th Cartesian coordinate. The system (2.6) is referred to as a "system of conservation laws" (see e.g. [34]). The equations in it are said to be written in "conservation" or "divergence" form since they directly descend from the conservation principles (2.3) by applying the divergence theorem (in consideration of the arbitrariness of V) [88].

The vector $\mathbf{q}$ defined in (2.4), commonly referred to as the "conservative" state vector, is chosen as the independent state vector.[1] Clearly, u_k can be recast as follows:

$$u_k = \frac{\hat{\mathbf{e}}^{(k)T} \cdot (\rho \mathbf{u})}{\rho} \tag{2.7}$$

and therefore $\mathbf{f}^{(k)}$ admits the following representation as a function of $\mathbf{q}$:

$$\mathbf{f}^{(k)}(\mathbf{q}) = \frac{\hat{\mathbf{e}}^{(k)T} \cdot (\rho \mathbf{u})}{\rho} \mathbf{q} + p(\rho) \begin{pmatrix} 0 \\ \hat{\mathbf{e}}^{(k)} \end{pmatrix}.$$

It is of interest to explicitly compute the following Jacobian:

$$\mathbf{J}^{(k)} := \partial_{\mathbf{q}} \mathbf{f}^{(k)}(\mathbf{q}) \tag{2.8}$$

in view of the fact that, for smooth solutions, the system (2.6) is equivalent to the following first-order quasi-linear one [55]:

$$\partial_t \, \mathbf{q} + \sum_{k=1}^{3} \mathbf{J}^{(k)} \cdot \partial_{x_k} \mathbf{q} = \mathbf{0}. \tag{2.9}$$

Then, the following expression is obtained by deriving (2.5) (by virtue of the relevant definitions introduced in Section 2.1):

$$\mathbf{J}^{(k)} = \mathbf{q} \cdot \partial_{\mathbf{q}} u_k + u_k \, \partial_{\mathbf{q}} \mathbf{q} + \begin{pmatrix} 0 \\ \hat{\mathbf{e}}^{(k)} \end{pmatrix} \cdot \partial_{\mathbf{q}} p. \tag{2.10}$$

---

[1] As pointed out, the "conservation" character of the system (2.6) is connected with its mathematical structure and it is by no means due to the specific choice of the "conservative" state vector as dependent variable.

By recalling (1.2)-(1.4), the derivative $\partial_q p$ is given by:

$$\partial_q p = \left(\begin{array}{c} a^2 \\ \mathbf{0} \end{array} \right)^T$$

where a denotes the sound speed and therefore:

$$\left(\begin{array}{c} 0 \\ \hat{\mathbf{e}}^{(k)} \end{array} \right) \cdot \partial_q p = \left(\begin{array}{cc} 0 & \mathbf{0}^T \\ a^2 \hat{\mathbf{e}}^{(k)} & \mathbf{O} \end{array} \right).$$

Moreover, by differentiating (2.7), the following relation is obtained:

$$\partial_q u_k = \rho^{-1} \left(\begin{array}{c} -u_k \\ \hat{\mathbf{e}}^{(k)} \end{array} \right)^T$$

and thus ($\mathbf{q}$ can be easily divided by ρ):

$$\mathbf{q} \cdot \partial_q u_k = \left(\begin{array}{c} 1 \\ \mathbf{u} \end{array} \right) \cdot \left(\begin{array}{c} -u_k \\ \hat{\mathbf{e}}^{(k)} \end{array} \right)^T = \left(\begin{array}{cc} -u_k & \hat{\mathbf{e}}^{(k)T} \\ -u_k \mathbf{u} & \mathbf{u} \cdot \hat{\mathbf{e}}^{(k)T} \end{array} \right).$$

Finally, $\partial_q \mathbf{q}$ is trivially equal to the identity matrix. Then, by exploiting the usual compact notation, it is possible to write the following equality:

$$u_k \, \partial_q \mathbf{q} = \left(\begin{array}{cc} u_k & \mathbf{0}^T \\ \mathbf{0} & u_k \mathbf{I} \end{array} \right).$$

By substituting the relevant entities into (2.10), the following representation is finally obtained:

$$\mathbf{J}^{(k)} = \left(\begin{array}{cc} 0 & \hat{\mathbf{e}}^{(k)T} \\ a^2 \hat{\mathbf{e}}^{(k)} - u_k \mathbf{u} & \mathbf{u} \cdot \hat{\mathbf{e}}^{(k)T} + u_k \mathbf{I} \end{array} \right). \tag{2.11}$$

Note 2.2.1. It is well-known that the system (2.3), or its differential counterpart (2.6), does not explicitly involve any similarity parameter [88]. This means that the considered governing equations can be thought to involve either dimensional or non-dimensional entities (flow, space and time variables). In the latter case, the non-dimensional form is obtained from the dimensional one by a standard technique [88], once introduced the following reference entities:

- x_{ref}: a reference length;
- u_{ref}: a reference speed;
- ρ_{ref}: a reference density;

- $t_{ref} := x_{ref} u_{ref}^{-1}$: a reference time;
- $p_{ref} := \rho_{ref} u_{ref}^2$: a reference pressure.

Of course, this observation immediately applies to any subsequent system of equations which is derived from (2.3) or (2.6) by means of simplifying assumptions. In particular, it directly applies to the Riemann problem introduced in Section 2.4 (the similarity character of its solution being preserved by the aforementioned non-dimensionalization procedure). Moreover, the considered observation holds true also when recasting the governing equations in a rotating frame (see Section 2.2.2) at the only cost of introducing an additional reference entity, namely:

- $\omega_{ref} := x_{ref}^{-1} u_{ref}$: a reference rotational speed.

2.2.2. 3D equations in rotating frames

With respect to a Cartesian frame having the same origin as that one introduced in Section 2.2.1 and rotating with constant angular velocity ω, the mass and momentum balances (2.3) read:

$$\partial_t \int_{\mathcal{V}} \mathbf{q} \, dV + \int_{\mathcal{S}} \left(\sum_{k=1}^{3} \hat{n}_k \, \mathbf{f}^{(k)} \right) dS = \int_{\mathcal{V}} \mathbf{s} \, dV \qquad (2.12)$$

with (relevant definitions from Section 2.2.1 are recalled):

$$\mathbf{s} := - \begin{pmatrix} 0 \\ 2\,\omega \wedge \rho \mathbf{u} + \rho\,\omega \wedge (\omega \wedge \mathbf{x}) \end{pmatrix} \qquad (2.13)$$

where $\mathbf{x}$ denotes the position of the generic fluid particle with respect to the considered rotating frame, $\mathbf{u}$ consistently represents the relative velocity and the symbol $\wedge$ indicates the usual vector product. The vector $\mathbf{s}$ accounts for the non-inertial effects related to the frame rotation; indeed, the terms $2\,\omega \wedge \rho \mathbf{u}$ and $\rho\,\omega \wedge (\omega \wedge \mathbf{x})$ in (2.13) respectively represent the analogue of the well-known Coriolis and centrifugal forces of classical rational mechanics [68]. Regular solutions of (2.12) also satisfy its differential counterpart, namely:

$$\partial_t \mathbf{q} + \sum_{k=1}^{3} \partial_{x_k} \mathbf{f}^{(k)} = \mathbf{s}. \qquad (2.14)$$

2.2.3. Basic-1D equations

Let u be the unique component of the velocity vector in a purely 1D motion along a certain direction, associated with the coordinate x; moreover, let $[\alpha, \beta]$ denote an arbitrary control volume along the x-axis. Once defined the 1D counterparts of (2.4) and (2.5) as follows:

$$\mathbf{q}^{(x)} := \begin{pmatrix} \rho \\ \rho u \end{pmatrix} \tag{2.15}$$

$$\mathbf{f}^{(x)} := \begin{pmatrix} \rho u \\ \rho u^2 + p \end{pmatrix} \tag{2.16}$$

the 1D balances corresponding to (2.3) and (2.6) respectively read:

$$\partial_t \int_\alpha^\beta \mathbf{q}^{(x)} \, dx + \mathbf{f}^{(x)}|_\beta - \mathbf{f}^{(x)}|_\alpha = \mathbf{0} \tag{2.17}$$

$$\partial_t \mathbf{q}^{(x)} + \partial_x \mathbf{f}^{(x)} = \mathbf{0} . \tag{2.18}$$

2.2.4. Augmented-1D equations

It is possible to extend the balances (2.17) and (2.18) so as to also describe the mass conservation of a certain substance merely advected with the flow (commonly referred to as a "passive scalar"). Indeed, once extended the definitions (2.15) and (2.16) as follows:

$$\mathbf{q}^{(A)} := \begin{pmatrix} \rho \\ \rho u \\ \rho \xi \end{pmatrix} \tag{2.19}$$

$$\mathbf{f}^{(A)} := \begin{pmatrix} \rho u \\ \rho u^2 + p \\ \rho u \xi \end{pmatrix} \tag{2.20}$$

where ξ denotes the concentration of the passive scalar, the "augmented" versions of the systems (2.17) and (2.18) respectively read (see *e.g.* [98]):

$$\partial_t \int_\alpha^\beta \mathbf{q}^{(A)} \, dx + \mathbf{f}^{(A)}|_\beta - \mathbf{f}^{(A)}|_\alpha = \mathbf{0} \tag{2.21}$$

$$\partial_t \mathbf{q}^{(A)} + \partial_x \mathbf{f}^{(A)} = \mathbf{0} . \tag{2.22}$$

Note 2.2.2. Clearly, the conservation of the passive scalar is decoupled from the basic 1D system (as, for instance, the energy balance which has been deliberately dropped out at the beginning of Section 2.2 for the sake of simplicity). Nevertheless, it is added to the basic 1D system in order to prepare the ground for the introduction of the 1D sweeps of the original 3D governing equations (see Section 2.2.5).

Note 2.2.3. It is straightforward to extend the "augmentation" procedure described in the present section to the case of m passive scalars, with $m > 1$. For instance, let ξ and η be two passive scalars; it is possible to formally keep (2.21) and (2.22) as they are, at the only cost of extending (2.19) and (2.20) as follows:

$$\mathbf{q}^{(A)} := \begin{pmatrix} \rho \\ \rho u \\ \rho \xi \\ \rho \eta \end{pmatrix} \qquad (2.23)$$

$$\mathbf{f}^{(A)} := \begin{pmatrix} \rho u \\ \rho u^2 + p \\ \rho u \xi \\ \rho u \eta \end{pmatrix}. \qquad (2.24)$$

It should be noticed that the m passive scalars, while not affecting the basic 1D flow field (*i.e.* ρ and u), neither interact with one another even. Therefore, the structure of the solution is identical for all of them, differences only arising due to the specific initial and boundary conditions. In view of this, it is reasonable to study the case $m = 1$, for the sake of simplicity.

2.2.5. 1D sweeps of the 3D equations

The "k-th sweep" of the 3D system (2.6) is obtained by neglecting the summation in it, namely (see *e.g.* [98]):

$$\partial_t \mathbf{q} + \partial_{x_k} \mathbf{f}^{(k)} = \mathbf{0} \quad , \quad k \in \{1, 2, 3\}. \qquad (2.25)$$

By respectively comparing the definitions of $\mathbf{q}$ and $\mathbf{f}^{(k)}$ in (2.4) and (2.5) with those of $\mathbf{q}^{(A)}$ and $\mathbf{f}^{(A)}$ in (2.23) and (2.24), it is clear that (apart from the order of the components) the k-th sweep (2.25) is formally equal to the augmented-1D system (2.22), at the cost of considering:

- the k-th coordinate direction (*i.e.* $\hat{\mathbf{e}}^{(k)}$) as the direction along which a basic-1D flow takes place;
- the velocity components u_h, with $h \in \{1, 2, 3\}$ and $h \neq k$, as advected passive scalars.

This observation is exploited in Section 5.1.2 in order to discretize the surface integral appearing in the 3D balances (2.3) and (2.12).

2.2.6. Hierarchical structure of the presented equations

The numerical discretization of the 3D problems (2.3) and (2.12) is discussed in Chapter 5. The rotating case, in particular, is treated as a generalization of the non-rotating one and therefore the problem (2.3) is considered at a preliminary stage.

Currently, a good mathematical understanding of the problems (2.3) and (2.6) is largely unavailable [34]. For this reason, the numerical discretization of (2.3) is based on some numerical techniques which are applied to the 1D sweeps of the original 3D problem. As mentioned in Section 2.2.5, the 1D sweeps are regarded to as augmented-1D systems and therefore the balances (2.21) and (2.22) are considered, in particular, for developing most of the proposed 1D numerical ingredients (see Chapter 3).

The basic-1D systems (2.17) and (2.18) are exploited in Section 3.4 to tackle some difficulties which are essentially related to the numerical discretization of the mass and momentum balances appearing in every augmented-1D system.

2.3. Hyperbolicity and integral solutions

In Sections 2.3.1 and 2.3.2, the hyperbolic character of the relevant governing equations introduced in Section 2.2 is concisely discussed. The concept of integral solution is then presented in Section 2.3.3 (together with some related issues like the Rankine-Hugoniot condition and the notion of entropy condition), in order to pave the way for discussing shock waves and contact discontinuities in Sections 2.4 and 2.5.

2.3.1. Hyperbolicity of the 3D equations

Let $\mathbf{J}$ represent the following matrix:

$$\mathbf{J} := \sum_{k=1}^{3} \hat{n}_k \, \mathbf{J}^{(k)} \qquad (2.26)$$

where $\hat{n}_k$ denotes the k-th component of the normal $\hat{\mathbf{n}}$ appearing in the 3D balances (2.3) and $\mathbf{J}^{(k)}$ is defined in (2.8). Once substituted the expression of $\mathbf{J}^{(k)}$ provided in (2.11), the following representation is obtained:

$$\mathbf{J} = \begin{pmatrix} 0 & \hat{\mathbf{n}}^T \\ a^2 \, \hat{\mathbf{n}} - \left(\hat{\mathbf{n}}^T \cdot \mathbf{u} \right) \mathbf{u} & \mathbf{u} \cdot \hat{\mathbf{n}}^T + \left(\hat{\mathbf{n}}^T \cdot \mathbf{u} \right) \mathbf{I} \end{pmatrix}. \qquad (2.27)$$

The quasi-linear system (2.9) is said to be hyperbolic (at a certain point of the flow field) if $\mathbf{J}$ has real eigenvalues λ_j and a corresponding set of linearly independent (right) eigenvectors $\mathbf{r}_j$ ($j = 1, \ldots, 4$), for every versor $\hat{\mathbf{n}}$ on the unit sphere $\mathbb{S}^2$. Furthermore, the system is said to be "strictly hyperbolic" if the eigenvalues are all distinct [55].

It is straightforward to verify that the system (2.9) is actually hyperbolic, with eigenvalues (the eigenvectors are not reported for the sake of conciseness):

$$\lambda_1 = \hat{\mathbf{n}} \cdot \mathbf{u} - a \quad , \quad \lambda_2 = \lambda_3 = \hat{\mathbf{n}} \cdot \mathbf{u} \quad , \quad \lambda_4 = \hat{\mathbf{n}} \cdot \mathbf{u} + a . \tag{2.28}$$

Note 2.3.1. The following equation:

$$\det (\mathbf{J} - \lambda \mathbf{I}) = 0 \tag{2.29}$$

with $\mathbf{J}$ given by (2.26), can be regarded to as a partial differential equation where the unknown is a certain function $z = z(x_1, x_2, x_3, t)$ such that:

$$\hat{\mathbf{n}} = \frac{\mathbf{z}}{\|\mathbf{z}\|} \quad , \quad \lambda = -\frac{\partial_t z}{\|\mathbf{z}\|}$$

where $\mathbf{z}$ denotes the (spatial) gradient of z. Manifolds $z = const$, where z is an integral solutions of (2.29), are "characteristic manifolds" having normal $\hat{\mathbf{n}}$ and moving with a normal component of the velocity equal to λ:

$$\lambda = \hat{\mathbf{n}} \cdot \frac{d\mathbf{x}}{dt} \tag{2.30}$$

where $\mathbf{x}$ here denotes the position of the generic point on the manifold (see *e.g.* [19, 20, 55, 56, 65] and [115]). By comparing (2.28) and (2.30), it is clear that the manifolds associated with λ_2 and λ_3 for the system (2.9) are simply advected with the flow (*i.e.* they behave like material surfaces) while those associated with λ_1 and λ_4 isotropically propagate along $\hat{\mathbf{n}}$ with the sound speed a.

Characteristic manifolds associated with the speeds (2.28) can transport discontinuities of the derivatives of the solution of (2.9) within the flow field (see *e.g.* [65]). This point, together with the well-known result that discontinuities can arise during the evolution of the solution also by starting from smooth data (see *e.g.* [34, 55] and [60]), clearly shows that it is not possible, in general, to find smooth solutions of the differential problem (2.6). Therefore, some way to interpret less regular solutions somehow "solving" (2.6), or its simplified counterpart (2.22), must be devised (see Section 2.3.3).

2.3.2. Hyperbolicity of the augmented-1D equations

The quasi-linear form of the system (2.22) reads:

$$\partial_t \, \mathbf{q}^{(A)} + \mathbf{J}^{(A)} \cdot \partial_x \mathbf{q}^{(A)} = 0 \qquad (2.31)$$

where $\mathbf{J}^{(A)}$ is the Jacobian of the function $\mathbf{f}^{(A)} \left(\mathbf{q}^{(A)} \right)$ defined by (2.19)-(2.20):

$$\mathbf{J}^{(A)} := \partial_{\mathbf{q}^{(A)}} \mathbf{f}^{(A)}. \qquad (2.32)$$

By analogy with the 3D case discussed in Section 2.3.1, the hyperbolicity of the system (2.31) depends on the eigenstructure of $\mathbf{J}^{(A)}$. In particular, it is straightforward to verify that it is strictly hyperbolic, with the following pairs of eigenvalue-eigenvector:

$$
\begin{cases}
\lambda_1 = u - a \quad , \quad \mathbf{r}_1 = (1, u - a, \xi)^T \\[2mm]
\lambda_2 = u \qquad , \quad \mathbf{r}_2 = (0, 0, 1)^T \\[2mm]
\lambda_3 = u + a \quad , \quad \mathbf{r}_3 = (1, u + a, \xi)^T .
\end{cases}
\qquad (2.33)
$$

2.3.3. Integral solutions

By following [34], a certain field $\mathbf{z} \in L^\infty \left(\mathbb{R} \times (0, \infty) \, ; \, \mathbb{R}^m \right)$ is said to be an integral solution of the following initial-value problem:

$$
\begin{cases}
\partial_t \, \mathbf{z} + \partial_x \mathbf{f} \; = \; 0 \quad \text{in} \quad \mathbb{R} \times (0, \infty) \\[2mm]
\mathbf{z} \; = \; \mathbf{z}^{(0)} \quad \text{on} \quad \mathbb{R} \times \{t = 0\}
\end{cases}
\qquad (2.34)
$$

with $\mathbf{f} = \mathbf{f}(\mathbf{z})$, once provided the following equality:

$$\int_0^\infty \int_{-\infty}^\infty (\mathbf{z} \cdot \partial_t \mathbf{v} + \mathbf{f} \cdot \partial_x \mathbf{v}) \; dx \, dt + \int_{-\infty}^\infty \mathbf{z}^{(0)} \cdot \mathbf{v}|_{t=0} \, dx = 0 \qquad (2.35)$$

holds for all the test functions $\mathbf{v}$ such that $\mathbf{v}$ is smooth and has compact support.

Note 2.3.2. The relation (2.35) is obtained integrating by parts the dot product between the p.d.e. in (2.34) and $\mathbf{v}$. Even if (2.35) is obtained by assuming that $\mathbf{z}$ is a smooth solution of (2.34), it makes sense if $\mathbf{z}$ is merely bounded. In consideration of the fact that the solution set of (2.35) contains that one of (2.34), the integral solutions are also called "weak" or "generalized" solutions (see *e.g.* [63] and [98]).

Rankine-Hugoniot condition

Let $\{(x, t) \mid x = s(t)\}$, for some smooth function $s(\cdot) : [0, \infty) \rightarrow \mathbb{R}$, represent a curve γ dividing a certain domain within $\mathbb{R} \times (0, \infty)$ into a "left" and a "right" sub-domain. Let $\mathbf{z}$ be an integral solution of (2.34) which is smooth on either sides of γ, along which $\mathbf{z}$ has simple jump discontinuities. Then, the integral solution must verify the classical Rankine-Hugoniot (hereafter RH as well) condition across γ (see *e.g.* [34]):

$$[\mathbf{f}] = \sigma \, [\mathbf{z}] \qquad (2.36)$$

where $\sigma := ds/dt$ and $[\psi]$ denotes the difference between the "left" and "right" limits of ψ across γ (or vice-versa, consistently on both sides of (2.36)). By analogy with classical fluid mechanics, the discontinuity along γ is commonly referred to as a "shock wave" or, briefly, "shock".

Entropy conditions

It is well known that integral solutions need not be unique and additional requirements for properly defining generalized solutions of (2.34) must be introduced [34]. A key issue, in particular, is the definition of "admissible shocks", *i.e.* discontinuities subjected to (2.36) which link a certain state $\mathbf{z}_2$ to a given state $\mathbf{z}_1$ in such a way that the evolution from $\mathbf{z}_1$ towards $\mathbf{z}_2$ is acceptable from a certain, say "physical", point of view while the reciprocal path is not. Criteria for selecting admissible shocks are called entropy conditions by analogy with classical gas dynamics, where the admissible shocks (from supersonic to subsonic regimes) are selected by exploiting the second principle of thermodynamics (*i.e.* the non-decreasing trend of the thermodynamic entropy) [66]. While allowing for the identification of the relevant physical evolution, the entropy conditions generally permit to obtain a unique solution of the mathematical problem.

Classical criteria like the Lax entropy condition [61] or the Liu entropy criterion (see *e.g.* [34]) provide restrictions on a possible couple of states joined by a shock. However, it is possible to widen the entropy criteria so as to apply to more general integral solutions of the considered conservation laws. In particular, it is possible to define so-called entropy solutions, *i.e.* solutions obeying certain requirements of the type of the Oleinik condition [72] (see *e.g.* [34]). The fundamental idea upon which the aforementioned entropy criteria are based is that physically and mathematically correct solutions of the p.d.e. in (2.34) should arise as the limit of the solutions to the following parabolic system (which admits travelling wave solutions, see *e.g.* [34]):

$$\partial_t \tilde{\mathbf{z}} + \partial_x \mathbf{f}\left(\tilde{\mathbf{z}}\right) = \varepsilon \; \partial_x \left(\partial_x \tilde{\mathbf{z}}\right) \quad , \quad \varepsilon > 0 \tag{2.37}$$

as the "viscosity" term on the right hand side vanishes (*i.e.* for $\varepsilon \to 0$).[2]

Another approach exploited for selecting relevant integral solutions consists in introducing suitable functions, called entropy functions, for which an additional conservation law holds for smooth solutions that becomes an inequality for discontinuous solutions (see *e.g.* [34] and [63]).

There is great ongoing interest in studying entropy conditions. Indeed, it is very difficult to assess criteria holding for general conservation laws, and in particular for generic relations $\mathbf{f}(\mathbf{z})$;[3] a vast number of results is currently available only for simplified systems or for scalar conservation laws [34]. In consideration of this, the classical Lax entropy condition is adopted (see Section 2.4.2), for the sake of simplicity, in order to determine the solution of the so-called Riemann problem (see Section 2.4) involving the system (2.22), closed by a convex barotropic state law (see Section 2.5.1).

2.4. The Riemann problem

In this section, the following system of conservation laws is considered:

$$\begin{cases} \partial_t \mathbf{z} + \partial_x \mathbf{f} = \mathbf{0} & \text{in} \quad \mathbb{R} \times (0, \infty) \\[2mm] \mathbf{z} = \begin{cases} \mathbf{z}_L & if \quad x < 0 \\ \mathbf{z}_R & if \quad x > 0 \end{cases} & \text{on} \quad \mathbb{R} \times \{t = 0\} \end{cases} \tag{2.38}$$

with $\mathbf{z} \in \mathbb{R}^m$ and $\mathbf{f} = \mathbf{f}(\mathbf{z}) \in \mathbb{R}^m$. This system, characterized by a step-like piece-wise constant initial data, is commonly referred to as the Riemann problem (hereafter RP as well).

The solution to (2.38) also depends, in general, on the state law closing the relevant p.d.e. through the specific relation $\mathbf{f}(\mathbf{z})$. This solution (when available, since for sophisticated state laws it is very difficult to be obtained) plays an important role in the set up of modern numerical methods for fluid dynamics (see *e.g.* [63, 98] and [99]) and therefore a vast class of RPs has been studied within this context, even involving complex state laws (see *e.g.* [13] and [69] amongst many others).

---

[2] For this reason, an entropy solution is also called a vanishing-viscosity solution.

[3] The specific form of the state law which closes the considered differential problem clearly affects $\mathbf{f}(\mathbf{z})$; in general, it can render it very difficult to define suitable entropy criteria. (see *e.g.* [3] and the cited references).

Once introduced some relevant definitions and results in Section 2.4.1, the basic wave solutions of (2.38) are presented in Section 2.4.2 and, finally, an important theorem related to the local solution of (2.38) is mentioned in Section 2.4.3. All the material concisely presented in Sections 2.4.1 to 2.4.3, essentially taken from [34] and [55], is aimed at preparing the ground for the solution of the specific RP studied in sec 2.5.

2.4.1. Preliminary definitions and results

The system in (2.38) is supposed to be strictly hyperbolic, with pairs $(\lambda_k, \mathbf{r}_k)$ $(k = 1, \ldots, m)$ of eigenvalue-eigenvector associated with the Jacobian $\partial_{\mathbf{z}}\mathbf{f}$ appearing in its quasi-linear form:

$$\partial_t \mathbf{z} + \partial_{\mathbf{z}}\mathbf{f} \cdot \partial_x \mathbf{z} = \mathbf{0}.$$

Characteristics

The following differential equations:

$$\frac{\mathrm{d}x}{\mathrm{d}\alpha} = \lambda_k(\mathbf{z}) \quad , \quad \frac{\mathrm{d}t}{\mathrm{d}\alpha} = 1 \tag{2.39}$$

where α is an abscissa, define the k-th family of characteristic curves (briefly: the k-th characteristics) in the $x - t$ plane, associated with the hyperbolic system in (2.38).

Genuinely non-linear and linearly-degenerate pairs

The pair $(\lambda_k, \mathbf{r}_k)$, with $\lambda_k = \lambda_k(\mathbf{z})$ and $\mathbf{r}_k = \mathbf{r}_k(\mathbf{z})$ is called genuinely non-linear (briefly: g.n.) provided:

$$\partial_{\mathbf{z}}\lambda_k \cdot \mathbf{r}_k \neq 0 \quad , \quad \forall \mathbf{z} \in \mathbb{R}^m.$$

Conversely, it is said to be linearly-degenerate (briefly: l.d.) if:

$$\partial_{\mathbf{z}}\lambda_k \cdot \mathbf{r}_k = 0 \quad , \quad \forall \mathbf{z} \in \mathbb{R}^m.$$

Rarefaction curves

Given a fixed state $\mathbf{z}_0 \in \mathbb{R}^m$, the k-th rarefaction curve $R_k(\mathbf{z}_0)$ is defined as the path in $\mathbb{R}^m$ of the solution to the following ordinary differential equation (hereafter o.d.e. as well):

$$\frac{\mathrm{d}\mathbf{v}(\xi)}{\mathrm{d}\xi} = \mathbf{r}_k(\mathbf{v}(\xi)) \tag{2.40}$$

which passes through $\mathbf{z}_0$. If $(\lambda_k, \mathbf{r}_k)$ is g.n., then (2.40) shows that λ_k monotonically increases or decreases along $R_k(\mathbf{z}_0)$ and therefore:

$$R_k(\mathbf{z}_0) = R_k^+(\mathbf{z}_0) \cup \{\mathbf{z}_0\} \cup R_k^-(\mathbf{z}_0)$$

with:

$$\begin{cases} R_k^+(\mathbf{z}_0) := \{\mathbf{z} \in R_k(\mathbf{z}_0) \mid \lambda_k(\mathbf{z}_0) < \lambda_k(\mathbf{z})\} \\ \\ R_k^-(\mathbf{z}_0) := \{\mathbf{z} \in R_k(\mathbf{z}_0) \mid \lambda_k(\mathbf{z}) < \lambda_k(\mathbf{z}_0)\} \,. \end{cases} \qquad (2.41)$$

Simple waves

A simple wave is a solution of the p.d.e. in (2.38) having the following structure:

$$\mathbf{z}(x, t) = \mathbf{v}(\eta(x, t)) \quad \text{in} \quad \mathbb{R} \times (0, \infty)\,. \qquad (2.42)$$

It is possible to show that $\mathbf{v}$ in (2.42) necessarily satisfies (2.40) for some k. In addition, η in (2.42) must satisfy the following p.d.e.:

$$\partial_t \eta + \lambda_k(\mathbf{v}(\eta)) \, \partial_x \eta = 0\,. \qquad (2.43)$$

The simple wave $\mathbf{z}$ given by (2.42) is consequently called a k-simple wave.

Shock set

Given a fixed state $\mathbf{z}_0 \in \mathbb{R}^m$, the so-called shock set is defined as follows:

$$S(\mathbf{z}_0) := \left\{\mathbf{z} \in \mathbb{R}^m \mid \mathbf{f}(\mathbf{z}) - \mathbf{f}(\mathbf{z}_0) = \sigma \, (\mathbf{z} - \mathbf{z}_0)\right\} \qquad (2.44)$$

where σ depends on the states $\mathbf{z}$ and $\mathbf{z}_0$: $\sigma = \sigma(\mathbf{z}, \mathbf{z}_0)$. It is possible to show that, in some neighbourhood of $\mathbf{z}_0$, $S(\mathbf{z}_0)$ consists of the union of m smooth curves $S_k(\mathbf{z}_0)$ ($k = 1, \ldots, m$) with the following properties:

- $S_k(\mathbf{z}_0)$ passes through $\mathbf{z}_0$ with tangent $\mathbf{r}_k(\mathbf{z}_0)$;
- $\sigma(\mathbf{z} \in S_k(\mathbf{z}_0), \mathbf{z}_0) = \dfrac{\lambda_k(\mathbf{z}) + \lambda_k(\mathbf{z}_0)}{2} + O\left(|\mathbf{z} - \mathbf{z}_0|^2\right)$ as $\mathbf{z} \to \mathbf{z}_0$.

Furthermore, it is possible to show that, if $(\lambda_k, \mathbf{r}_k)$ is g.n., then (provided $\mathbf{z}$ is close enough to $\mathbf{z}_0$):

$$S_k(\mathbf{z}_0) = S_k^+(\mathbf{z}_0) \cup \{\mathbf{z}_0\} \cup S_k^-(\mathbf{z}_0) \qquad (2.45)$$

with:

$$\begin{cases} S_k^+(\mathbf{z}_0) := \{\mathbf{z} \in S_k(\mathbf{z}_0) \mid \lambda_k(\mathbf{z}_0) < \sigma(\mathbf{z}, \mathbf{z}_0) < \lambda_k(\mathbf{z})\} \\ \\ S_k^-(\mathbf{z}_0) := \{\mathbf{z} \in S_k(\mathbf{z}_0) \mid \lambda_k(\mathbf{z}) < \sigma(\mathbf{z}, \mathbf{z}_0) < \lambda_k(\mathbf{z}_0)\} \,. \end{cases} \qquad (2.46)$$

Linear degeneracy

If $(\lambda_k, \mathbf{r}_k)$ is l.d. for some $k \in \{1, \ldots, m\}$, then it is possible to show that, for each $\mathbf{z}_0 \in \mathbb{R}^m$:

- $S_k(\mathbf{z}_0) = R_k(\mathbf{z}_0)$;
- $\sigma(\mathbf{z}, \mathbf{z}_0) = \lambda_k(\mathbf{z}) = \lambda_k(\mathbf{z}_0)$, $\forall \mathbf{z} \in S_k(\mathbf{z}_0)$.

Note 2.4.1. It should be noticed that the curves $S_k(\mathbf{z}_0)$ and $R_k(\mathbf{z}_0)$, which in general agree at least to the first order at $\mathbf{z}_0$, coincide in the linearly-degenerate case.

2.4.2. Basic wave solutions

The basic wave solutions of the RP (2.38) are considered in the sequel.

Rarefactions

It is possible to show that there exists a continuous integral solution $\mathbf{z}$ of the RP (2.38) which is a k-simple wave constant along lines [4] through the origin ($x = 0, t = 0$), for $\lambda_k(\mathbf{z}_L) \leq x/t \leq \lambda_k(\mathbf{z}_R)$, provided that:

- $(\lambda_k, \mathbf{r}_k)$ is g.n.;
- $\mathbf{z}_R \in R_k^+(\mathbf{z}_L)$.

For this solution, $\eta(x, t)$ in (2.42) is given by $\eta(x, t) = \bar{\eta}(x/t)$, for a suitable function $\bar{\eta}$. Since η is constant along a k-th characteristic by virtue of (2.43) and (2.39), it follows that also the k-th characteristics, for $\lambda_k(\mathbf{z}_L) \leq x/t \leq \lambda_k(\mathbf{z}_R)$, are lines through the origin and therefore (2.39), in particular, becomes:

$$\frac{\mathrm{d}x}{\mathrm{d}t} = \frac{x}{t} = \lambda_k(\mathbf{z}) \quad , \quad \lambda_k(\mathbf{z}_L) \leq \frac{x}{t} \leq \lambda_k(\mathbf{z}_R). \tag{2.47}$$

The k-th characteristics for the solution under consideration, also referred to as a "k-rarefaction (wave)" by analogy with classical gas dynamics [66], are sketched in Figure 2.1.

Note 2.4.2. The relation (2.40), in particular, holds for the solution at hand; once recast in differential form, it states that the solution $\mathbf{z} \in \mathbb{R}^m$, while moving between two infinitesimally close lines within the wave region sketched in Figure 2.1, must satisfy the following condition:

$$\mathrm{d}\mathbf{z} \propto \mathbf{r}_k(\mathbf{z}) \tag{2.48}$$

---

[4] In the present document, the word "line" denotes a straight curve while the word "curve" stands for a generic curve.

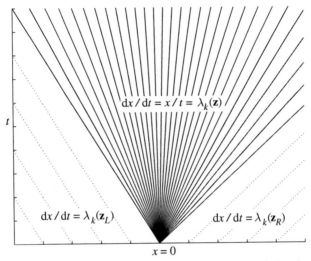

Figure 2.1. Schematic representation of the k-th characteristics for a generic k-rarefaction.

The above condition provides $m - 1$ independent scalar differential relations which can be integrated, thus providing $m - 1$ prime integrals across the wave.[5] The prime integrals under consideration are sometimes called generalized Riemann invariants (see *e.g.* [55] and [98]).[6]

Shocks - Lax entropy condition

If $\mathbf{z}_R \in S_k(\mathbf{z}_L)$, the following field:

$$\mathbf{z}(x, t) = \begin{cases} \mathbf{z}_L & \text{if } x < \sigma t \\ \mathbf{z}_R & \text{if } x > \sigma t \end{cases} \tag{2.49}$$

with $\sigma = \sigma(\mathbf{z}_R, \mathbf{z}_L)$, is an integral solution of (2.38). By virtue of (2.44), the solution at hand satisfies the RH condition (2.36) and it is consequently referred to as a shock (wave).

If $(\lambda_k, \mathbf{r}_k)$ is g.n. then $\mathbf{z}_R$ (provided it is close enough to $\mathbf{z}_L$) can belong either to $S_k^+(\mathbf{z}_L)$ or to $S_k^-(\mathbf{z}_L)$, due to (2.45). By adopting the classical Lax entropy condition [61] (hereafter LEC as well), only the latter possibility is considered acceptable; once recalled the definition of S_k^- in (2.46), it is possible to recast the LEC as follows:

$$\lambda_k(\mathbf{z}_R) < \sigma(\mathbf{z}_R, \mathbf{z}_L) < \lambda_k(\mathbf{z}_L). \tag{2.50}$$

---

[5] Since $\mathbf{r}_k$ is only determined up to an arbitrary multiplicative constant, it might be necessary to choose the multiplicative factor tacitly appearing in (2.48) so as to be a suitable integrating factor.

[6] Indeed they generalize the classical Riemann invariants, which do not exist, in general, for $m > 2$ (see *e.g.* [34]).

Then, the shock (2.49) is accepted as an integral solution of (2.38) if and only if the pair $(\mathbf{z}_R, \mathbf{z}_L)$ satisfies (2.50).

Note 2.4.3. According to the LEC (2.50), the k-th characteristics from left and right (lines on both sides) run into the shock, as sketched in Figure 2.2. Since the characteristics act as information carriers (see *e.g.* [55] and [65]), some information is lost when they reach the shock, thus increasing a suitably defined entropy of the system (see *e.g.* [34]). This, in turn, is considered as a proper criterion for assessing the physical representativeness of the shock, in analogy with the case of classical gas dynamics in which the LEC paraphrases the non-decreasing character of the thermodynamic entropy, *i.e.* the second principle of thermodynamics [66]. In view of this consideration, it is clear why the LEC is regarded to as an entropy condition.

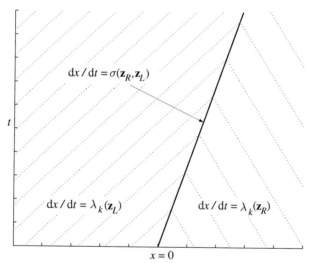

Figure 2.2. Schematic representation of the k-th characteristics for a generic shock satisfying the LEC (2.50).

Contact discontinuities

The expression (2.49) is an integral solution of (2.38) also when $(\lambda_k, \mathbf{r}_k)$ is l.d., at the obvious cost of choosing $\sigma = \sigma(\mathbf{z}_R, \mathbf{z}_L) = \lambda_k(\mathbf{z}_L) = \lambda_k(\mathbf{z}_R)$, as imposed by the linear degeneracy (see Section 2.4.1). The left and right k-th characteristics (lines on both sides) are then parallel to the discontinuity, as sketched in Figure 2.3. This solution is called a k-contact discontinuity.

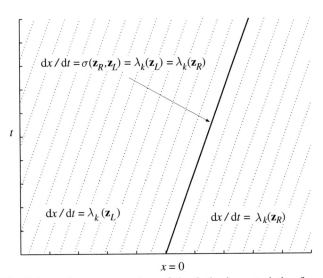

$x = 0$

Figure 2.3. Schematic representation of the k-th characteristics for a generic k-contact discontinuity.

Note 2.4.4. In consideration of the fact that $S_k(\mathbf{z}) = R_k(\mathbf{z})$ for the contact discontinuity, as imposed by the linear degeneracy (see Section 2.4.1), the jump $[\mathbf{z}]$ across the discontinuity can be computed by exploiting either the RH condition (2.36) or by integrating (2.48) (as for a rarefaction).

2.4.3. Local solution of the Riemann problem

An important theorem shows that, if $(\lambda_k, \mathbf{r}_k)$ is either g.n. or l.d. for each $k \in \{1, \dots, m\}$, then there exists an integral solution of the RP (2.38) which is constant on lines through the origin ($x = 0, t = 0$), provided the initial states $\mathbf{z}_L$ and $\mathbf{z}_R$ are sufficiently close to each other [34]. While proving the aforementioned statement, it is possible to construct the solution by connecting $m + 1$ states $\mathbf{z}_h$ ($h = 0, \dots, m$) by means of m waves of the type of those discussed in Section 2.4.2 (*i.e.* rarefactions, shocks and contact discontinuities). More in detail, let $\lambda_k < \lambda_{k+1}$ (which is legitimate, due to the strict hyperbolicity assumed at the beginning of Section 2.4.1); then, once chosen $\mathbf{z}_0 = \mathbf{z}_L$ and $\mathbf{z}_m = \mathbf{z}_R$, the wave joining $\mathbf{z}_{k-1}$ to $\mathbf{z}_k$ is a k-th rarefaction, a shock or a k-th contact discontinuity, provided:

$$\mathbf{z}_k \in T_k(\mathbf{z}_{k-1})$$

where the curve $T_k(\mathbf{z})$ is defined, in some neighbourhood of $\mathbf{z}$, as follows:

$$T_k(\mathbf{z}) := \begin{cases} R_k^+(\mathbf{z}) \cup \{\mathbf{z}\} \cup S_k^-(\mathbf{z}) & \text{if} \quad (\lambda_k, \mathbf{r}_k) \quad \text{g.n.} \\ \\ R_k(\mathbf{z}) = S_k(\mathbf{z}) & \text{if} \quad (\lambda_k, \mathbf{r}_k) \quad \text{l.d.} \end{cases} \qquad (2.51)$$

This constructive procedure is applied in Section 2.5.3 in order to solve the RP (2.38) when associated with a generic but convex barotropic state law.

2.5. The Riemann problem for a convex barotropic state law

In this section, the following RP is considered:

$$
\begin{cases}
\partial_t \mathbf{q}^{(A)} + \partial_x \mathbf{f}^{(A)} = \mathbf{0} & \text{in} \quad \mathbb{R} \times (0, \infty) \\[2ex]
\mathbf{q}^{(A)} = \begin{cases} \mathbf{q}_L^{(A)} & if \quad x < 0 \\ \mathbf{q}_R^{(A)} & if \quad x > 0 \end{cases} & \text{on} \quad \mathbb{R} \times \{t = 0\}
\end{cases}
\tag{2.52}
$$

with $\mathbf{q}^{(A)}$ and $\mathbf{f}^{(A)}$ given by (2.19) and (2.20), respectively. The p.d.e. in (2.52) (*i.e.* the system (2.22)) is supposed to be closed by a generic state law like that one defined in Section 1.5, subjected to an additional constraint that is discussed in Section 2.5.1. In Section 2.5.2 basic wave solutions are investigated, which are exploited in Section 2.5.3 for constructing the solution of the RP (2.52).

2.5.1. Convexity of the state law

In [69] general constitutive relations involving several thermodynamic entities are investigated and the convexity of a given state law is defined within a quite general context. To the purposes of the present work, it suffices to mention that the generic barotropic state law (1.3) is said to be convex if:

$$
\frac{d^2 p}{dv^2} > 0
\tag{2.53}
$$

where $v := \rho^{-1}$ is the so-called specific volume. As a function of ρ, the term on the left-hand side of the condition (2.53) can be recast as follows:

$$
\frac{d^2 p}{dv^2} = 2a\,\rho^4\,c(\rho)
\tag{2.54}
$$

where:

$$
c(\rho) := \frac{a}{\rho} + \frac{da}{d\rho}.
\tag{2.55}
$$

Hence, the condition (2.53) can be equivalently expressed as follows:

$$
c(\rho) > 0.
\tag{2.56}
$$

Some considerations can be drawn from the convexity condition (2.56), namely:

- let $\chi = \chi(\rho)$ be defined as follows:

$$\chi(\rho) := \rho \, a(\rho). \tag{2.57}$$

Clearly $d\chi/d\rho = \rho \, c(\rho)$ and therefore for a convex barotropic state law χ is a monotonically increasing function of ρ;
- let $\Theta(\rho_0, \rho)$ be defined as follows (a prolongation by continuity is considered):

$$\Theta(\rho_0, \rho) := \begin{cases} \rho_0 \, \rho \, \dfrac{p - p_0}{\rho - \rho_0} & \text{if } \rho \neq \rho_0 \\[2mm] \chi_0^2 & \text{if } \rho = \rho_0 \end{cases} \tag{2.58}$$

where $\chi_0 := \chi(\rho_0)$, with χ defined in (2.57). In consideration of the stability constraint (1.4), $\Theta > 0$. Moreover, for a convex barotropic state law it is possible to show that:

$$\begin{cases} \Theta(\rho_0, \rho) > \chi_0^2 & \text{if } \rho > \rho_0 \\[2mm] \Theta(\rho_0, \rho) < \chi_0^2 & \text{if } \rho < \rho_0. \end{cases} \tag{2.59}$$

To the purpose, it suffices to consider the following function:

$$\Gamma(\rho) := (\rho - \rho_0)\left(\Theta - \chi_0^2\right)$$

in which ρ_0 is regarded to as a fixed parameter. Indeed, by virtue of the convexity condition (2.56), $d^2\Gamma/d\rho^2 > 0$ while $d\Gamma/d\rho = 0$ for $\rho = \rho_0$. Hence, Γ has one and only one minimum in correspondence of $\rho = \rho_0$; since $\Gamma(\rho = \rho_0) = 0$, it follows that $\Gamma > 0$ for $\rho \neq \rho_0$ and therefore $\left(\Theta - \chi_0^2\right)$ has the same sign as $(\rho - \rho_0)$ (thus obtaining (2.59)).

Once introduced the following definition:

$$\zeta(\rho_0, \rho) := +\Theta(\rho_0, \rho)^{\frac{1}{2}} \tag{2.60}$$

with Θ given by (2.58), it is possible to recast the inequalities (2.59) (which only involve positive entities) as follows:

$$\begin{cases} \zeta(\rho_0, \rho) > \chi_0 & \text{if } \rho > \rho_0 \\[2mm] \zeta(\rho_0, \rho) < \chi_0 & \text{if } \rho < \rho_0. \end{cases} \tag{2.61}$$

Then, in consideration of the symmetry $\zeta\,(\rho_0, \rho) = \zeta\,(\rho, \rho_0)$ and by inverting the role of ρ_0 and ρ in (2.61), it follows that for a convex barotropic state law:

$$\chi_0 < \zeta\,(\rho_0, \rho) = \zeta\,(\rho, \rho_0) < \chi \quad , \qquad \rho_0 < \rho \tag{2.62}$$

where, of course, $\chi = \chi(\rho)$ according to (2.57);
- let $\Phi = \Phi(\rho)$ be defined as follows:

$$\Phi(\rho) := \Psi(\rho) + a(\rho) \tag{2.63}$$

where:

$$\Psi(\rho) := \int_{\rho_0}^{\rho} \frac{a(s)}{s}\,\mathrm{d}s\,. \tag{2.64}$$

Clearly $\mathrm{d}\Phi/\mathrm{d}\rho = c(\rho)$ and therefore for a convex barotropic state law Φ is a monotonically increasing function of ρ.

Contrarily to the thermodynamic stability constraint (1.4), the convexity condition (2.56) is not imposed by physical requirements. Nevertheless, it is attractive from a mathematical point of view since it makes it possible to construct a local solution to the RP (2.52) by juxtaposing the wave solutions introduced in Section 2.4.2 (*i.e.* rarefactions, shocks and contact discontinuities). Indeed, the convexity condition (2.56) renders all the pairs $(\lambda_k, \mathbf{r}_k)$ $(k \in \{1, 2, 3\})$ defined in (2.33) either g.n. or l.d., since:

$$\begin{cases} \partial_{\mathbf{q}^{(A)}}\lambda_1 \cdot \mathbf{r}_1 &= -c(\rho) < 0 \quad \text{g.n.} \\[2mm] \partial_{\mathbf{q}^{(A)}}\lambda_2 \cdot \mathbf{r}_2 &= 0 \qquad\qquad \text{l.d.} \\[2mm] \partial_{\mathbf{q}^{(A)}}\lambda_3 \cdot \mathbf{r}_3 &= +c(\rho) > 0 \quad \text{g.n.} \end{cases} \tag{2.65}$$

and thus permits to exploit the results reported in Section 2.4.3. A generic convex barotropic state law is therefore assumed in Sections 2.5.2 and 2.5.3, in order to construct the solution to the RP (2.52).

Note 2.5.1. As an example, it is straightforward to verify that the following law:

$$p = p_{\text{model}}(\rho) := \kappa\,\rho^\chi + \gamma \tag{2.66}$$

with $\kappa > 0$, $\varkappa > 0$ and γ given (real) constants, is convex. It should be noticed that:

- for $\gamma = 0$ the classical polytropic gas state law is obtained;
- for $\kappa = 2^{-1}$, $\varkappa = 2$ and $\gamma = 0$ the conservation laws (2.22) become formally identical to the well-known homogeneous shallow water equations (of course, augmented by the advection of the passive scalar), for which there exists a vast literature also investigating RPs (see *e.g.* [99]);
- for $\kappa = \varepsilon\,\rho_0^{-\varkappa}$ and $\gamma = -\varepsilon$, with $\varepsilon > 0$ and $\rho_0 > 0$ given constants, the classical Tait law (which is used for describing isentropic compressible liquids) is formally recovered; the corresponding RP is studied, for instance, in [51].

Note 2.5.2. The rheological behaviour of a large variety of real-world materials cannot be represented by convex laws and therefore non-convex state laws have been studied as well, for incorporation into classical systems of equations for fluid dynamics (see *e.g.* [69, 113] and [114]). The solution of the RP associated with a non-convex state law is, in general, more difficult than that one associated with a convex one since it admits, besides the basic waves discussed in Section 2.4.2, more complex wave solutions (see the aforementioned references).

2.5.2. Basic wave solutions

As already noticed in Section 2.3.2, the system (2.22) is strictly hyperbolic, with pairs of eigenvalue-eigenvector given by (2.33). In this section, basic k-waves (*i.e.* wave solutions associated with the pair $(\lambda_k, \mathbf{r}_k)$) are investigated, of the type of those discussed in Section 2.4.2. In view of the material presented in the aforementioned section, the relation (2.65) clearly implies that the 2-waves are necessarily contact discontinuities while the others can be either rarefactions or shocks.

In the rest of this section, the considered waves are supposed to separate a "left" state $\mathbf{q}_l^{(A)}$ and a "right" state $\mathbf{q}_r^{(A)}$. The subscripts l and r are also exploited for concisely representing entities related to the aforementioned states.

1-rarefaction

Let $q_h^{(A)}$ ($h \in \{1, 2, 3\}$) denote the h-th component of $\mathbf{q}^{(A)}$. Once recalled the eigenstructure (2.33) of the system at hand, the definition of the generalized Riemann invariants (2.48) leads to the following differential relations:

$$\begin{cases} d\left(\dfrac{q_2^{(A)}}{q_1^{(A)}}\right) + \dfrac{a(q_1^{(A)})}{q_1^{(A)}} dq_1^{(A)} = 0 \\[3mm] \dfrac{dq_3^{(A)}}{q_3^{(A)}} - \dfrac{dq_1^{(A)}}{q_1^{(A)}} = 0 \end{cases}$$

which integrate to:

$$\begin{cases} u_r - u_l = \Psi_l - \Psi_r \\[2mm] \xi_r - \xi_l = 0 \end{cases} \tag{2.67}$$

where Ψ is given by (2.64).

Proposition 2.5.3. *The rarefaction under consideration is a wave solution of the RP (2.52) if and only if:*

$$\rho_r < \rho_l \tag{2.68}$$

Proof. Since the pair $(\lambda_1, \mathbf{r}_1)$ is g.n., it is necessary and sufficient for the wave under consideration to be a solution that (see the relevant paragraph in Section 2.4.2):

$$\mathbf{q}_r^{(A)} \in R_1^+\left(\mathbf{q}_l^{(A)}\right).$$

For the present case, the aforementioned condition reads $\lambda_{1l} < \lambda_{1r}$ or, equivalently, $a_r - a_l < u_r - u_l$. Then, by substituting the first relation in (2.67), the inequality under consideration can be recast as follows:

$$\Phi_r < \Phi_l \tag{2.69}$$

with Φ defined in (2.63). By recalling the fact that, for a convex barotropic state law, $\Phi(\rho)$ is a monotonically increasing function, the conditions (2.68) and (2.69) are equivalent to each other. This concludes the proof. $\qquad\square$

1-shock

By introducing the relevant definitions into the RH condition (2.36), the following relations are obtained:

$$\begin{cases} \rho_r \bar{u}_r - \rho_l \bar{u}_l = 0 \\[2mm] \rho_r \bar{u}_r^2 - \rho_l \bar{u}_l^2 = p_l - p_r \\[2mm] \xi_r - \xi_l = 0 \end{cases} \tag{2.70}$$

where:

$$\bar{u}_j := u_j - \sigma \quad , \quad j \in \{l, r\} \tag{2.71}$$

and $\sigma = \sigma(u_l, u_r)$ denotes the shock speed.[7] By manipulating the first two equations in (2.70) the following relations are obtained:

$$\rho_l \bar{u}_l = \rho_r \bar{u}_r = \zeta(\rho_l, \rho_r) \tag{2.72}$$

$$u_r - u_l = \frac{\rho_l - \rho_r}{\rho_l \rho_r} \zeta(\rho_l, \rho_r) \tag{2.73}$$

with ζ given by (2.60). Moreover, by substituting (2.72) into (2.71), the following expression is obtained for the shock speed σ:

$$\sigma = u_j - \rho_j^{-1} \zeta(\rho_l, \rho_r) \quad , \quad j \in \{l, r\}.$$

Proposition 2.5.4. *The shock under consideration is a wave solution of the RP (2.52) if and only if:*

$$\rho_l < \rho_r. \tag{2.74}$$

Proof. Since the pair $(\lambda_1, \mathbf{r}_1)$ is g.n., the LEC (2.50) is a necessary and sufficient condition for the shock under consideration to be admissible (see the relevant paragraphs in Section 2.4.2). By exploiting (2.72), the LEC (2.50) can be recast as follows:

$$\chi_l < \zeta(\rho_l, \rho_r) < \chi_r \tag{2.75}$$

with χ defined in (2.57). Then, by recalling (2.62), it is clear that for a convex barotropic state law the condition (2.75) (*i.e.* the LEC (2.50)) and the condition (2.74) are equivalent to each other. This concludes the proof. □

2-contact discontinuity

Let $q_h^{(A)}$ ($h \in \{1, 2, 3\}$) denote the h-th component of $\mathbf{q}^{(A)}$. By applying (2.48) to the present case (see Note 2.4.4 in Section 2.4.2), the following differential relations are obtained:

$$\begin{cases} dq_1^{(A)} = 0 \\ dq_2^{(A)} = 0 \end{cases}$$

---

[7] The first relation in (2.70), directly derived from the first component of (2.36), is exploited to obtain the representation of the others.

which trivially integrate to:

$$\begin{cases} \rho_r - \rho_l = 0 \\ u_r - u_l = 0. \end{cases} \qquad (2.76)$$

Moreover, the speed σ of the contact discontinuity is straightforwardly given by (see Section 2.4.2):

$$\sigma = u_l = u_r$$

3-rarefaction

Once noticed that $\mathbf{r}_3$ reduces to $\mathbf{r}_1$ by inverting the sign of the sound speed a, (2.67) directly implies that across the waves under consideration the following relations hold:

$$\begin{cases} u_r - u_l = \Psi_r - \Psi_l \\ \xi_r - \xi_l = 0 \end{cases} \qquad (2.77)$$

where Ψ is given by (2.64).

Proposition 2.5.5. *The rarefaction under consideration is a wave solution of the RP (2.52) if and only if:*

$$\rho_l < \rho_r . \qquad (2.78)$$

Proof. Analogous to the that one of Proposition 2.5.3 above.   □

3-shock

By introducing the relevant definitions into the RH condition (2.36), the following relations are obtained (identical to those in (2.70)):

$$\begin{cases} \rho_r \bar{u}_r - \rho_l \bar{u}_l = 0 \\ \rho_r \bar{u}_r^2 - \rho_l \bar{u}_l^2 = p_l - p_r \\ \xi_r - \xi_l = 0 \end{cases} \qquad (2.79)$$

where $\bar{u}_j$ ($j \in \{l, r\}$) is defined in (2.71). By manipulating the first two equations in (2.79) the following relations are obtained:

$$\rho_l \bar{u}_l = \rho_r \bar{u}_r = -\zeta (\rho_l, \rho_r) \qquad (2.80)$$

$$u_r - u_l = \frac{\rho_r - \rho_l}{\rho_l \rho_r} \zeta (\rho_l, \rho_r) \qquad (2.81)$$

with ζ given by (2.60).[8] Moreover, by substituting (2.80) into (2.71), the following expression is obtained for the shock speed σ:

$$\sigma = u_j + \rho_j^{-1} \zeta (\rho_l, \rho_r) \quad , \quad j \in \{l, r\}. \tag{2.82}$$

Proposition 2.5.6. *The shock under consideration is a wave solution of the RP (2.52) if and only if:*

$$\rho_r < \rho_l . \tag{2.83}$$

Proof. Analogous to the that one of Proposition 2.5.4 above. $\qquad\square$

2.5.3. Local solution of the Riemann problem

As remarked in Section 2.5.1, by adopting a convex barotropic state law it is possible to exploit the constructive procedure outlined in Section 2.4.3 in order to define a local solution of the RP (2.52). Moreover, since two generic adjacent states appearing in the solution are only connected to each other by means of a basic wave (*i.e.* a rarefaction, a shock or a contact discontinuity), it is possible to use the relations obtained in Section 2.5.2, as explained below. The solution strategy outlined in this section is then exploited in Chapters 3 and 4 in order to validate 1D numerical methods.

Structure of the solution

By recalling the theorem mentioned in Section 2.4.3, it is clear that the solution of the RP (2.52) in general consists of three waves of the type of those discussed in Section 2.5.2. These waves separate four states, $\mathbf{q}_L^{(A)}$, $\mathbf{q}_{L\star}^{(A)}$, $\mathbf{q}_{R\star}^{(A)}$ and $\mathbf{q}_R^{(A)}$, amongst which $\mathbf{q}_L^{(A)}$ and $\mathbf{q}_R^{(A)}$ are given by the initial condition associated with the RP (2.52) while the others must be determined. The structure of the wave solution is sketched in Figure 2.4; in particular, the solid line represents the 2-contact discontinuity while each couple of dotted lines denotes either a shock or a rarefaction.

In consideration of (2.67), (2.70), (2.76), (2.77) and (2.79) it is evident that ξ only varies across the contact discontinuity while ρ and u (which are continuous across the contact discontinuity) change, in general, across the other waves. Hence, the state vectors under consideration

---

[8] The difference between (2.72) and (2.80) arises from the LEC (2.50). Indeed, in the former case it must be (in particular) $\lambda_{1l} = u_l - a_l > \sigma$, i.e. $\bar{u}_l > a_l$ while in the latter one it must be (in particular) $\sigma > \lambda_{3r} = u_r + a_r$, i.e. $\bar{u}_r < -a_r$. Since the sound speed is always positive, in the former case $\rho_j \bar{u}_j = \pi_1$ while in the latter one $\rho_j \bar{u}_j = -\pi_3$, with π_1 and π_3 positive entities. Straightforward computations show that $\pi_1 = \pi_3 = \zeta (\rho_l, \rho_r)$.

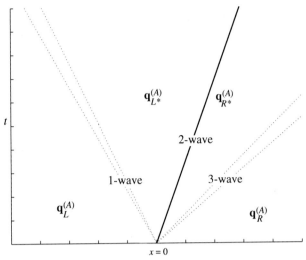

Figure 2.4. Schematic representation of the solution of the RP (2.52). The 2-wave is a contact discontinuity while the others can be either a shock or a rarefaction.

admit the following representation:

$$\mathbf{q}_L^{(A)} = \begin{pmatrix} \rho_L \\ \rho_L u_L \\ \rho_L \xi_L \end{pmatrix} \quad , \quad \mathbf{q}_{L\star}^{(A)} = \begin{pmatrix} \rho_\star \\ \rho_\star u_\star \\ \rho_\star \xi_L \end{pmatrix}$$

$$\mathbf{q}_{R\star}^{(A)} = \begin{pmatrix} \rho_\star \\ \rho_\star u_\star \\ \rho_\star \xi_R \end{pmatrix} \quad , \quad \mathbf{q}_R^{(A)} = \begin{pmatrix} \rho_R \\ \rho_R u_R \\ \rho_R \xi_R \end{pmatrix}$$

where $\rho_\star$ and $u_\star$ need to be defined in order to completely determine the solution.

Determination of the solution

By combining the expressions which give the variation of u across the waves, namely (2.67), (2.73), (2.77) and (2.81), it is straightforward to express $u_\star$ as a function of $\rho_\star$ as follows:

$$u_\star = u_L - \Omega\left(\rho_\star, \rho_L\right) = u_R + \Omega\left(\rho_\star, \rho_R\right) \qquad (2.84)$$

where (a prolongation by continuity for $\rho = \rho_j$ is considered):

$$\Omega\left(\rho, \rho_j\right) := \begin{cases} \Psi\left(\rho\right) - \Psi\left(\rho_j\right) & \text{if } \rho \le \rho_j \\[2mm] \dfrac{\rho - \rho_j}{\rho\,\rho_j} \zeta\left(\rho, \rho_j\right) & \text{if } \rho > \rho_j \end{cases} \quad , \quad j \in \{L, R\}$$

with Ψ and ζ respectively given by (2.64) and (2.60). Let Δu be defined as follows:

$$\Delta u := u_R - u_L .$$

Then, in view of (2.84), the identity $(u_L - u_\star)+(u_\star - u_R)+(u_R - u_L) = 0$ can be recast as follows:

$$\Omega\,(\rho_\star, \rho_L) + \Omega\,(\rho_\star, \rho_R) + \Delta u = 0 . \qquad (2.85)$$

Clearly, for the relation (2.85) to hold, $\rho_\star$ must be a zero of the following function:

$$\Omega_{(L,R)}\,(\rho) := \Omega_L\,(\rho) + \Omega_R\,(\rho) + \Delta u \qquad (2.86)$$

where:

$$\Omega_j\,(\rho) := \Omega\left(\rho, \rho_j\right) \quad , \quad j \in \{L, R\} .$$

The existence and the uniqueness of such a zero is ensured by the following:

Proposition 2.5.7. *Let $D_\rho = [\rho_{\min}, \rho_{\sup})$ denote the density domain of the considered barotropic state law, as defined in Section 1.5. There exists a unique solution $\rho_\star \in D_\rho$ of the equation (2.85) if and only if:*

$$\Delta^{\inf}u < \Delta u \le \Delta^{\max}u \qquad (2.87)$$

with:

$$\begin{cases} \Delta^{\inf}u & := \quad -\dfrac{\rho_{\sup} - \rho_L}{\rho_{\sup}\,\rho_L}\,\zeta\left(\rho_{\sup}, \rho_L\right) - \dfrac{\rho_{\sup} - \rho_R}{\rho_{\sup}\,\rho_R}\,\zeta\left(\rho_{\sup}, \rho_R\right) \\[4mm] \Delta^{\max}u & := \quad \Psi\,(\rho_L) + \Psi\,(\rho_R) - 2\,\Psi\,(\rho_{\min}) . \end{cases}$$

Proof. The first derivative of $\Omega_j\,(\rho)$ is given by the following (continuous) function:

$$\frac{d}{d\rho}\,\Omega_j\,(\rho) = \begin{cases} \dfrac{a}{\rho} & \text{if} \quad \rho \le \rho_j \\[4mm] \dfrac{1}{2}\left(\dfrac{\zeta\,(\rho, \rho_j)}{\rho^2} + \dfrac{a^2}{\zeta\,(\rho, \rho_j)}\right) & \text{if} \quad \rho > \rho_j \end{cases} \quad , \; j \in \{L, R\}$$

which is clearly positive. Hence, $\Omega_{(L,R)}\,(\rho)$ in (2.86) is a monotonically increasing function and admits a unique zero, which moves towards lower values of the density as Δu increases. By continuity, there exists a maximum value of Δu, denoted by $\Delta^{\max}u$, for which $\rho_\star = \rho_{\min}$ as well as an inferior one, denoted by $\Delta^{\text{inf}} u$, in correspondence of which $\rho_\star = \rho_{\text{sup}}$. Clearly, $\Delta^{\text{max}} u$ can be determined by evaluating (2.85) in correspondence of $\rho = \rho_{\text{min}}$. In particular, since $\rho_{\text{min}} \leq \rho_j$, $j \in \{L, R\}$, it follows (from the definitions) that:

$$\Omega_L (\rho_{\text{min}}) + \Omega_R (\rho_{\text{min}}) = 2 \Psi (\rho_{\text{min}}) - \Psi (\rho_L) - \Psi (\rho_R)$$

and therefore (2.85) in the present case reads:

$$2 \Psi (\rho_{\text{min}}) - \Psi (\rho_L) - \Psi (\rho_R) + \Delta^{\text{max}} u = 0$$

Similar considerations can be exploited for deriving the expression of $\Delta^{\text{inf}} u$. It is evident that (2.87) represents a necessary and sufficient condition for determining a solution $\rho_\star \in D_\rho$ of the non-linear equation (2.85). This completes the proof.   □

Note 2.5.8. As Δu transitions between $\Delta^{\text{inf}} u$ and $\Delta^{\text{max}} u$, the wave structure of the solution of the RP (2.52) changes. Once introduced the following definitions:

$$\begin{cases} \Delta^{2s} u & := \quad -\dfrac{\rho_M - \rho_m}{\rho_M \rho_m} \zeta (\rho_M, \rho_m) \\[2ex] \Delta^{2r} u & := \quad \Psi (\rho_M) - \Psi (\rho_m) \end{cases}$$

where:

$$\rho_M := \max(\rho_L, \rho_R) \quad , \quad \rho_m := \min(\rho_L, \rho_R)$$

and by recalling the monotonicity of $\Omega_{(L,R)}$, it is possible to identify the following sequence of wave solutions:

- for $\Delta^{\text{inf}} u < \Delta u < \Delta^{2s} u$, $\rho_M < \rho_\star < \rho_{\text{sup}}$ and both the 1-wave and the 3-wave are shocks;
- for $\Delta^{2s} u \leq \Delta u < \Delta^{2r} u$, $\rho_m < \rho_\star \leq \rho_M$; there is a shock between $\rho_\star$ and ρ_m, and a rarefaction between $\rho_\star$ and ρ_M;
- for $\Delta^{2r} u \leq \Delta u \leq \Delta^{\text{max}} u$, $\rho_{\text{min}} \leq \rho_\star \leq \rho_m$ and both the 1-wave and the 3-wave are rarefactions.

The aforementioned statements can be straightforwardly verified by introducing considerations of the same kind of those reported in the proof of Proposition 2.5.7 (above) for determining $\Delta^{\text{max}} u$.

Note 2.5.9. The solution for ξ depends on $\rho_\star$ and $u_\star$ (since $u_\star$ determines the location of the contact discontinuity) but, in turns, it does not affect the solution for ρ and u. This point, which is due to the decoupling between the passive scalar and the underlying 1D flow field (see Note 2.2.2

in Section 2.2.4), permits to straightforwardly extend the structure of the considered solution to the case of an arbitrary number $m > 1$ of passive scalars. Indeed, since there is no interaction between them (see Note 2.2.3 in Section 2.2.4), it suffices to make all of them simultaneously jump across the contact discontinuity. Even if for $m > 1$ the system (2.22) ceases to be strictly hyperbolic (the multiplicity of the eigenvalue associated with the contact discontinuity being in general equal to m), the fact that the additional waves do not interact with the starting system, neither with one another even, makes the loss of strict hyperbolicity purely formal. In other words, the augmented system with $m > 1$ behaves like that one having $m = 1$ and it is possible to keep the proposed solution strategy.

Note 2.5.10. Clearly, the solution of the considered RP is essentially constructed by solving (2.85). It is therefore evident that it is possible to keep the proposed solution procedure also for formulations adopting p instead of ρ as the independent variable, at the only cost of straightforward changes in the notation.

Note 2.5.11. It is worth mentioning that the material presented in Sections 2.5.2 and 2.5.3 generalizes the solution procedure reported in [99] for the RP associated with the homogeneous shallow water equations (see Note 2.5.1 in Section 2.5.1).

Chapter 3
1D Numerical method

A 1D numerical method (hereafter also referred to as numerical scheme) for discretizing the augmented-1D equations (2.22) is developed in the present section. In particular, the integral form of the considered conservation law, namely (2.21), is considered in order to allow integral solutions (see Section 2.3.3) to be taken into account. A shock-capturing approach (see *e.g.* [39, 64, 80] and [98]) is chosen, in order to approximate possibly discontinuous solutions.

Once introduced some basic material in Section 3.1, a Godunov scheme for (generic) convex barotropic state laws is proposed in Section 3.2. A Roe scheme is then proposed in Section 3.3, which can be applied when dealing with generic barotropic state laws. In Section 3.4 the behaviour of this scheme in the nearly-incompressible limit is investigated and a suitable preconditioning technique for low Mach number flows is introduced. Finally, in Section 3.5 a linearization of a generic Roe numerical flux function is proposed, only relying on its algebraic properties and therefore applicable to a variety of problems. The proposed linearization is then applied to the barotropic case under consideration, in order to define a linearized implicit time-advancing strategy.

3.1. Generalities on the 1D discretization

In the sequel, some basic concepts and definitions which are related to the numerical discretization of the considered 1D conservation law are introduced, to be exploited within the rest of Chapter 3. The concise introduction under consideration does not lay claim to yield a rigorous and complete treatment of the subject; a detailed presentation can be found in a number of textbooks (*e.g.* [39, 64] and [98] amongst many others).

3.1.1. Space discretization

A finite volume approach is adopted for the spatial discretization of the problem (2.22). The x-domain is divided into N_c cells, indexed by $i \in$

$\mathcal{I} := \{1, \ldots, N_c\}$. The i-th cell spans the interval $C_i := (x_{i-1/2}, x_{i+1/2})$, with $x_{i-1/2} < x_{i+1/2}$, having measure μ_i. On C_i the exact solution $\mathbf{q}^{(A)}(x, t)$ is approximated by a semi-discrete function $\mathbf{q}_i^{(A)}(t)$, which is considered as an approximation of the mean value of $\mathbf{q}^{(A)}(x, t)$ over C_i:

$$\mathbf{q}_i^{(A)}(t) \approx \frac{1}{\mu_i} \int_{C_i} \mathbf{q}^{(A)}(x, t) \, dx \, . \tag{3.1}$$

The differential system defining $\mathbf{q}_i^{(A)}$ is obtained by discretizing the integral balance (2.21) over the control volume C_i. Indeed, by virtue of (3.1), the time-derivative in (2.21) is naturally approximated as follows:

$$\frac{\partial}{\partial t} \int_{C_i} \mathbf{q}^{(A)}(x, t) \, dx \approx \mu_i \frac{d}{dt} \mathbf{q}_i^{(A)} \tag{3.2}$$

while the inter-cell flux $\mathbf{f}^{(A)}$ defined in (2.20) is approximated by introducing a suitable numerical flux function (hereafter numerical flux, as well) $\boldsymbol{\phi}^{(A)}$, depending on the semi-discrete solution.

Numerical flux

Let π_i denote the set of indexes identifying the cells in the neighbourhood of C_i, namely:

$$\pi_i := \{i - 1, i + 1\} \, . \tag{3.3}$$

A certain degree of locality is usually assumed for $\boldsymbol{\phi}^{(A)}$ and the flux crossing the boundary between C_i and C_j towards C_j is commonly approximated by means of the following expression:

$$\boldsymbol{\phi}^{(A)} \left(\mathbf{q}_i^{(A)}, \mathbf{q}_j^{(A)}, \hat{\boldsymbol{v}}_{ij} \right) \, , \quad j \in \pi_i \tag{3.4}$$

where $\hat{\boldsymbol{v}}_{ij}$ is the versor associated with a generic vector mapping an arbitrary point of C_i to an arbitrary point of C_j. If $\hat{\boldsymbol{e}}$ denotes the versor associated with the x-axis, then clearly:

$$\hat{\boldsymbol{v}}_{ij} = (j - i) \hat{\boldsymbol{e}} \, , \quad j \in \pi_i \, . \tag{3.5}$$

The following definition is consequently introduced, to be exploited in the sequel:

$$s_{ij} := \hat{\boldsymbol{v}}_{ij} \cdot \hat{\boldsymbol{e}} = \text{sign}(j - i) \, , \quad j \in \pi_i \, . \tag{3.6}$$

Note 3.1.1. The explicit dependence of $\boldsymbol{\phi}^{(A)}$ on $\hat{\boldsymbol{v}}_{ij}$ in (3.4) may seem somewhat redundant in the present context. Indeed, a 1D case is intrinsically structured: any internal cell C_i has two and only two neighbours,

C_{i-1} and C_{i+1}, and $\hat{v}_{ij} = \pm \hat{e}$ according to (3.5). However, the proposed formulation (3.4), allows for an extension to 2D and 3D -possibly unstructured- spatial discretizations to be obtained (see *e.g.* Section 5.1.2), since it does not *a priori* incorporate any structure.

In general, the numerical flux must satisfy the following basic requirements:

$$\phi^{(A)} \left(\mathbf{q}_j^{(A)}, \mathbf{q}_i^{(A)}, \hat{v}_{ji} = -\hat{v}_{ij} \right) = - \phi^{(A)} \left(\mathbf{q}_i^{(A)}, \mathbf{q}_j^{(A)}, \hat{v}_{ij} \right) \qquad (3.7)$$

$$\phi^{(A)} \left(\mathbf{q}_i^{(A)}, \mathbf{q}_j^{(A)} = \mathbf{q}_i^{(A)}, \hat{v}_{ij} \right) = s_{ij} \; \mathbf{f}^{(A)} \left(\mathbf{q}_i^{(A)} \right). \qquad (3.8)$$

The property (3.7) is directly inherited from the continuous flux; it permits to associate the numerical flux with the inter-cell boundary in a well-defined way, thus allowing for the definition of "conservative" numerical schemes (see Note 3.1.2 in Section 3.1.2). The property (3.8), instead, enforces a natural consistency requirement.

Godunov approach

By following the well-known approach originally proposed by Godunov [40], the piece-wise constant approximant $\mathbf{q}_i^{(A)}$ ($i \in \mathcal{I}$) can be considered as defining a local RP like (2.52) at each interface $x_{(i+j)/2}$ ($j \in \pi_i$). The numerical flux between C_i and C_j can therefore be constructed by properly exploiting the solution either of the RP of interest (see Section 3.2.1) or of a suitable approximation of it (see Section 3.3.1).

Semi-discrete formulation and boundary conditions

By exploiting (3.2) and (3.4) it is straightforward to obtain the following class of semi-discrete approximations of (2.21), depending on the specific choice of the numerical flux:

$$\mu_i \frac{d}{dt} \mathbf{q}_i^{(A)} + \sum_{j \in \pi_i} \phi^{(A)} \left(\mathbf{q}_i^{(A)}, \mathbf{q}_j^{(A)}, \hat{v}_{ij} \right) = 0 \; , \; i \in \mathcal{I}. \qquad (3.9)$$

At the present stage of the discussion, the scheme (3.9) is not properly defined for $i = 1$ and $i = N_c$: suitable boundary conditions (hereafter BCs as well) must be introduced in order to completely define the semi-discrete formulation. To the purpose, two fictitious state vectors, $\mathbf{q}_0^{(A)}$ and $\mathbf{q}_{N_c+1}^{(A)}$, are introduced. Once these vectors have been given a value (modelling the chosen BCs), it is possible to directly apply (3.9) to every cell C_i ($i \in \mathcal{I}$).

3.1.2. Time discretization: basic discrete schemes

A fully-discrete (hereafter discrete) numerical scheme approximating (2.21) is defined by considering (3.9) as an ordinary differential equation.[1] The discrete solution at time-level $n + 1$ (corresponding to $t = t^{n+1}$), denoted by $\mathbf{q}_i^{(A)n+1}$ within cell C_i, can therefore be obtained from that one at time-level n by exploiting a variety of integration techniques. Basic discrete schemes are presented below.

Explicit time-advancing

An explicit discrete scheme can be obtained from (3.9) by considering, for instance, the classical "forward Euler" integration technique (see *e.g.* [79]):

$$\mathbf{q}_i^{(A)n+1} = \mathbf{q}_i^{(A)n} - \frac{\delta^n t}{\mu_i} \sum_{j \in \pi_i} \boldsymbol{\phi}^{(A)} \left(\mathbf{q}_i^{(A)n}, \mathbf{q}_j^{(A)n}, \hat{\mathbf{v}}_{ij} \right) , \ i \in \mathcal{I} \qquad (3.10)$$

where:

$$\delta^n(\cdot) := (\cdot)^{n+1} - (\cdot)^n . \qquad (3.11)$$

Note 3.1.2. Due to its specific form, the scheme (3.10) is said to be "conservative" (see *e.g.* [39] and [98]). It is very important to exploit conservative schemes in order to compute possibly discontinuous integral solutions; indeed, non-conservative formulations do not converge[2] to the correct solution of the problem if it involves shock waves [48]. A classical result, on the other hand, states that conservative numerical methods, if convergent, do converge to an integral solution of the considered conservation law [62]. Hence, it is practically compulsory to exploit conservative schemes when adopting a shock-capturing numerical approach.[3]

Implicit time-advancing

An implicit discrete scheme can be obtained from (3.9) approximating the time derivative by means of a backward finite difference as follows:

$$\frac{\mu_i}{\delta^n t} \delta^n \mathbf{q}_i^{(A)} + \sum_{j \in \pi_i} \boldsymbol{\phi}^{(A)} \left(\mathbf{q}_i^{(A)n+1}, \mathbf{q}_j^{(A)n+1}, \hat{\mathbf{v}}_{ij} \right) = \mathbf{0} , \ i \in \mathcal{I} . \qquad (3.12)$$

---

[1] This approach, keeping space and time discretizations separate, is sometimes referred to as a "method of lines" (see *e.g.* [39]).

[2] The notion of convergence, even if not formally introduced, is assumed to be understood at this point of the discussion.

[3] Different choices can be considered when adopting other techniques (*e.g.* shock-fitting or adaptive primitive-conservative numerical methods) [98].

It is well known that implicit schemes like (3.12) permit a more efficient time-advancing than explicit ones, because they do not suffer from time-step limitations caused by CFL-like stability constraints (see *e.g.* [44]). However, the scheme (3.12) can be demanding from a computational point of view, since it requires the solution of a non-linear system at each time-level. Indeed, the flux function is, in general, non-linear and the specific form of the considered state law can add to the complexity of the algorithm. As a matter of fact, 3D numerical schemes based on the extension of (3.12), are exceedingly intensive from a computational point of view[4] especially if they are applied to industrial problems involving very complex geometries.

Linearized implicit time-advancing

In view of the above considerations, a reasonable compromise between a purely explicit and a purely implicit scheme seems to be provided by a linearized implicit time-advancing strategy. This technique is based on the following approximate linearization of the numerical flux (which, in general, is not differentiable), assuming it exists:

$$\delta^n \phi_{ij}^{(A)} \approx \mathbf{A}_{ij}^{(A)n} \cdot \delta^n \mathbf{q}_i^{(A)} + \mathbf{B}_{ij}^{(A)n} \cdot \delta^n \mathbf{q}_j^{(A)} \tag{3.13}$$

where:

$$\phi_{ij}^{(A)n} := \phi^{(A)} \left(\mathbf{q}_i^{(A)n}, \mathbf{q}_j^{(A)n}, \hat{\mathbf{v}}_{ij} \right)$$

and:

$$\begin{cases} \mathbf{A}_{ij}^{(A)n} & := \mathbf{A}^{(A)} \left(\mathbf{q}_i^{(A)n}, \mathbf{q}_j^{(A)n}, \hat{\mathbf{v}}_{ij} \right) \\ \\ \mathbf{B}_{ij}^{(A)n} & := \mathbf{B}^{(A)} \left(\mathbf{q}_i^{(A)n}, \mathbf{q}_j^{(A)n}, \hat{\mathbf{v}}_{ij} \right) \end{cases} \tag{3.14}$$

with $\mathbf{A}^{(A)}$ and $\mathbf{B}^{(A)}$ suitably defined matrices. By substituting (3.13) into (3.12), the following scheme is obtained:

$$\left(\frac{\mu_i}{\delta^n t} \mathbf{I} + \sum_{j \in \pi_i} \mathbf{A}_{ij}^{(A)n} \right) \cdot \delta^n \mathbf{q}_i^{(A)} + \sum_{j \in \pi_i} \mathbf{B}_{ij}^{(A)n} \cdot \delta^n \mathbf{q}_j^{(A)} = - \sum_{j \in \pi_i} \phi_{ij}^{(A)n} , \quad i \in \mathcal{I}. \tag{3.15}$$

The scheme (3.15) represents a linear system (in particular, a block tridiagonal system in the considered 1D case) for the unknowns $\delta^n \mathbf{q}_i^{(A)}$; once it has been solved, the unknowns at time-level $n+1$ are trivially given by

---

[4] Unless exploiting specific supercomputing resources which are usually not available for common research or even industrial projects.

$q_i^{(A)n+1} = q_i^{(A)n} + \delta^n q_i^{(A)}$. Clearly, the linearized scheme (3.15) involves an additional degree of approximation with respect to the implicit scheme (3.12) (due to the approximate linearization[5]) but it is less demanding from a computational point of view. For this reason, a linearized implicit time-advancing is proposed in Section 3.5.

The scheme (3.15) can be regarded to as a particular instance of a more general linearized implicit formulation which is described below. The semi-discrete formulation (3.9) can be rewritten, in a more general way, as follows:

$$\frac{d}{dt} z_h + \psi_h^{(p)}(z_h) = 0 \tag{3.16}$$

where z_h denotes a suitable state vector representing the semi-discrete solution and $\psi_h^{(p)}(\cdot)$ denotes a vector operator whose components are spatial difference operators (the superscript p is discussed below). It is possible to approximate the time derivative as follows:

$$\frac{d}{dt} z_h(t^{n+1}) \approx \frac{\alpha_k z_h^{n+1} - z_h^{(n,k)}}{\delta^n t} \tag{3.17}$$

where $z_h^{(n,k)}$ denotes a linear combination of $z_h^n, \ldots, z_h^{n+1-k}$ and α_k is a scalar. Clearly, by suitably defining α_k and $z_h^{(n,k)}$ it is possible to obtain a certain order of accuracy (in the sense of the finite difference schemes) for the approximation of the time derivative. Once the approximation (3.17) has been substituted into the semi-discrete scheme (3.16), its discrete counterpart reads:

$$\mu_h^{(p,k)}(z_h^{n+1}) = 0 \tag{3.18}$$

where the operator $\mu_h^{(p,k)}(\cdot)$ is defined as follows:

$$\mu_h^{(p,k)}(\bar{z}_h) := \frac{\alpha_k}{\delta^n t} \bar{z}_h + \psi_h^{(p)}(\bar{z}_h) - \frac{1}{\delta^n t} z_h^{(n,k)}. \tag{3.19}$$

The non-linear discrete problem (3.18)-(3.19) can then be solved by means of a variety of techniques. In particular, it is possible to:

(i) introduce an approximate linearization of the operator $\mu_h^{(p,k)}$, as follows:

$$\mu_h^{(p,k)}(z_h^{n+1}) - \mu_h^{(p,k)}(z_h^n) \approx J_h^{(p,k)}(z_h^n) \cdot \delta^n z_h \tag{3.20}$$

---

[5] Of course, the effects the approximate linearization has on the numerical solution may be relatively less important for simulations marching towards a steady-state.

where:

$$\mathbf{J}_h^{(p,k)}\left(\cdot\right) := \frac{\alpha_k}{\delta^n t}\mathbf{I} + \delta\boldsymbol{\psi}_h^{(p)}\left(\cdot\right) \tag{3.21}$$

and $\delta\boldsymbol{\psi}_h^{(p)}\left(\cdot\right)$ denotes an approximation of the Jacobian of $\boldsymbol{\psi}_h^{(p)}\left(\cdot\right)$ or, more in general, a term rendering the approximation (3.20) acceptable. Once (3.20) has been substituted into (3.18), it is straightforward to solve the resulting linear problem with respect to $\delta^n\mathbf{z}_h$ (the discrete scheme (3.15), for instance, is obtained by following this approach). Of course, the computational cost of the considered strategy is mainly determined by the inversion of the linear operator $\mathbf{J}_h^{(p,k)}$, which can be still demanding for complex 3D industrial problems. When dealing with structured grids, for instance, it is possible to contain the computational cost by applying an approximate factorization technique to $\mathbf{J}_h^{(p,k)}$ [11]; indeed, the introduction of an additional discretization error due to the factorization may be acceptable, in consideration of the fact that simpler linear systems must be solved;

(ii) iteratively solve (3.18) by determining a fixed-point of the following relation, which implicitly maps $\mathbf{z}_h^{\lambda}$ to $\mathbf{z}_h^{\lambda+1}$ (with $\mathbf{z}_h^{\lambda=0} = \mathbf{z}_h^n$ as starting point:[6]

$$\boldsymbol{\mu}_h^{(q,k)}\left(\mathbf{z}_h^{\lambda+1}\right) = \boldsymbol{\mu}_h^{(q,k)}\left(\mathbf{z}_h^{\lambda}\right) - \boldsymbol{\mu}_h^{(p,k)}\left(\mathbf{z}_h^{\lambda}\right) \tag{3.22}$$

where $\boldsymbol{\mu}_h^{(q,k)}$ is formally defined by (3.19), with q in place of p. Once introduced the following linearization (in the spirit of (3.20)):

$$\boldsymbol{\mu}_h^{(q,k)}\left(\mathbf{z}_h^{\lambda+1}\right) - \boldsymbol{\mu}_h^{(q,k)}\left(\mathbf{z}_h^{\lambda}\right) \approx \mathbf{J}_h^{(q,k)}\left(\mathbf{z}_h^{\lambda}\right) \cdot \delta^{\lambda}\mathbf{z}_h$$

the map (3.22) can be explicitly approximated as follows:

$$\mathbf{z}_h^{\lambda+1} \approx \mathbf{z}_h^{\lambda} - \left(\mathbf{J}_h^{(q,k)}\left(\mathbf{z}_h^{\lambda}\right)\right)^{-1} \cdot \left(\boldsymbol{\mu}_h^{(p,k)}\left(\mathbf{z}_h^{\lambda}\right)\right). \tag{3.23}$$

Obviously, for the considered strategy to be computationally attractive with respect to that one described in the point (i) above, the inversion of $\mathbf{J}_h^{(q,k)}$ must be cheaper than that one of $\mathbf{J}_h^{(p,k)}$.[7] In particular, as for the point (i) above, it is possible to contain the considered computational cost when dealing with structured grids by applying an approximate factorization technique to $\mathbf{J}_h^{(q,k)}$ [11]. For practical purposes, the

---

[6] The considered iterations advance the solution with respect to the counter λ and the fixed-point $\mathbf{z}_h^{\lambda=\bar{\lambda}}$ represents the discrete solution at time-level $n + 1$ (i.e. $\mathbf{z}_h^{n+1}$) independently of the specific value of $\bar{\lambda}$. In order to emphasize this concept, the iterations at hand are sometimes referred to as "internal iterations" or "pseudo-iterations".

[7] Typically, $\mathbf{J}_h^{(q,k)}$ is sparser than $\mathbf{J}_h^{(p,k)}$.

solution is advanced only for a limited number, say λ_{max}^n, of iterations and the considered scheme globally reads:

$$
\begin{cases}
\mathbf{z}_h^{\lambda=0} = \mathbf{z}_h^n \\[2mm]
\mathbf{z}_h^{\lambda+1} = \mathbf{z}_h^\lambda - \left(\mathbf{J}_h^{(q,k)}\left(\mathbf{z}_h^\lambda\right)\right)^{-1} \cdot \left(\boldsymbol{\mu}_h^{(p,k)}\left(\mathbf{z}_h^\lambda\right)\right) , \quad \lambda=0,\ldots,\left(\lambda_{max}^n - 1\right) \\[2mm]
\mathbf{z}_h^{n+1} = \mathbf{z}_h^{\lambda=\lambda_{max}^n} .
\end{cases}
$$

$$(3.24)$$

If the superscripts p and q denote a formal order of accuracy of the corresponding spatial operators, it suffices, in general, to choose $q <$ p for making the inversion of $\mathbf{J}_h^{(q,k)}$ cheaper than that one of $\mathbf{J}_h^{(p,k)}$. Let k denote the formal order of accuracy of the discretization of the time derivative in (3.17); in particular, let $k = p$ for the sake of simplicity. Under these assumptions, the solution of the discrete scheme (3.18) as well as the fixed-point solution of (3.22) are formally of order p. On the other hand, the solution obtained by taking a single step of (3.24) is of order (q, p), *i.e.* of order $q < p$ for the space discretization and of order p for the time discretization. Nevertheless, it is possible to recover a p-order accuracy within a certain number of time-steps, without fully converging to the fixed-point solution. This consideration, which is at the basis of the "Defect Correction" methods (see *e.g.* [67]), renders the iterative scheme (3.24) appealing from a computational point of view. The definition of a suitable DeC scheme is mentioned in Section 3.5.3;

(iii) adopt a dual time-stepping approach [52], according to which the solution $\mathbf{z}_h^{n+1}$ in (3.18) is obtained by advancing the following problem:

$$
\begin{cases}
\dfrac{\mathrm{d}}{\mathrm{d}\tau}\,\mathbf{y}_h + \boldsymbol{\mu}_h^{(p,k)}\left(\mathbf{y}_h\right) &= \mathbf{0} \\[3mm]
\mathbf{y}_h\left(\tau = 0\right) &= \mathbf{z}_h^n
\end{cases}
$$

$$(3.25)$$

with respect to the pseudo-time τ, up to a steady-state. Of course, the numerical scheme specifically adopted for discretizing the pseudo-time derivative in (3.25) characterizes the considered "artificial" evolution between time-level n and time-level $n + 1$.

3.2. A Godunov scheme for convex barotropic state laws

In Section 3.2.1 a Godunov numerical flux function applicable to generic convex barotropic state laws is defined. In Section 3.2.2, the scheme

(3.10) exploiting the considered numerical flux is validated against an exact solution.

3.2.1. Godunov numerical flux

Let

$$\mathbf{q}_{RP}^{(A)}\left(\mathbf{q}_L^{(A)}, \mathbf{q}_R^{(A)}, \zeta\right)$$

denote the solution of a RP having "left" and "right" initial states $\mathbf{q}_L^{(A)}$ and $\mathbf{q}_R^{(A)}$, respectively, in correspondence of $x/t = \zeta$. The Godunov numerical flux at the interface $x_{(i+j)/2}$, with $j \in \pi_i$ and π_i given by (3.3), is constructed by evaluating the analytical flux $\mathbf{f}^{(A)}$ in correspondence of the following state vector:

$$\mathbf{q}_{RP}^{(A)}\left(\mathbf{q}_{L_{ij}}^{(A)}, \mathbf{q}_{R_{ij}}^{(A)}, 0\right)$$

where:

$$L_{ij} := \min(i, j) , \quad R_{ij} := \max(i, j)$$

and $\zeta = 0$ (*i.e.* $x = 0$) is chosen for correctly picking out the considered interface with respect to the local x-coordinate system to which the initial states are referred. The orientation defined by $\hat{\mathbf{v}}_{ij}$ is straightforwardly taken into account by defining the Godunov numerical flux as follows:

$$\boldsymbol{\phi}^{(A)\mathrm{GOD}}\left(\mathbf{q}_i^{(A)}, \mathbf{q}_j^{(A)}, \hat{\mathbf{v}}_{ij}\right) := s_{ij}\, \mathbf{f}^{(A)}\left(\mathbf{q}_{RP}^{(A)}\left(\mathbf{q}_{L_{ij}}^{(A)}, \mathbf{q}_{R_{ij}}^{(A)}, 0\right)\right), \quad j \in \pi_i \quad (3.26)$$

with s_{ij} given by (3.6).

Note 3.2.1. It should be noticed that the numerical flux (3.26) satisfies the conservation property (3.7); indeed, $L_{ji} = L_{ij}$ and $R_{ji} = R_{ij}$, while $s_{ji} = -s_{ij}$. The consistency property (3.8) is satisfied as well, by virtue of the consistency of $\mathbf{q}_{RP}^{(A)}(\cdot, \cdot, \cdot)$ which does not perturb a uniform (trivial) initial condition.

For the case of a generic convex barotropic state law (see Section 2.5.1), it is possible to define the numerical flux (3.26) by exploiting the solution $\mathbf{q}_{RP}^{(A)}(\cdot, \cdot, \cdot)$ proposed in Section 2.5.

3.2.2. Numerical results

The solution of a chosen RP is considered as a quantitative benchmark for validating the discrete scheme (3.10), based on the proposed numerical flux (3.26).

Benchmarks

The considered benchmarks are summarized in Table 3.1. In this table, κ, $\varkappa$ and γ refer to the chosen convex state law (2.66), ρ_L, u_L, ξ_L, ρ_R, u_R and ξ_R characterize the initial condition (hereafter IC as well) associated with the RP and t_{eval} denotes the time at which the considered solution is picked. The instances of the convex state law (2.66) incorporated in

Table 3.1. Considered benchmarks.

Benchmark	κ	$\varkappa$	γ	ρ_L	u_L	ξ_L	ρ_R	u_R	ξ_R	t_{eval}
B1	10^6	1	0	1.02	10	2	1	20	4	1
B2	10^6	2	0	1.02	10	2	1	20	4	1

Table 3.1 are simple power laws which permit to control the characteristic sound speed and the wave structure (hence, the flow compressibility) of the solution to the considered RP, by tuning the IC. In particular:

- for the state law considered in the benchmark B1, the sound speed is constant: $a = \sqrt{\kappa} = 10^3$ and therefore $\tilde{a} = 10^3$ represents the characteristic sound speed of the flow. Once chosen $\rho_R = 1$, $u_L = \tilde{M}\tilde{a}$ with $\tilde{M} = 10^{-2}$ and $u_R = 2u_L$, the value of ρ_L is tuned so as to obtain a left rarefaction and a right shock, with $u_*/\tilde{a} = O(\tilde{M})$ (in particular $\rho_L/\rho_R = 1 + O(\tilde{M})$). Hence, $\tilde{M} = 10^{-2}$ represents a characteristic Mach number for the whole flow field under consideration;
- for the state law considered in the benchmark B2, the sound speed varies with the density as follows: $a = \sqrt{2\kappa\rho}$. Once chosen $\rho_R = 1$, $u_L = \zeta a_R/\sqrt{2}$ with $\zeta = 10^{-2}$ and $u_R = 2u_L$, the value of ρ_L is tuned so as to obtain a left rarefaction and a right shock, with $u_*/a_R = O(\zeta)$ (in particular $\rho_L/\rho_R = 1 + O(\zeta)$). In the present case, the characteristic sound speed and the characteristic Mach number of the flow are $\tilde{a} = a_R$ and $\tilde{M} = \zeta$ (in particular, $\tilde{M} = 10^{-2}$ as for the benchmark B1).

Note 3.2.2. The structure of the solution to the considered RPs (*i.e.* rarefaction, contact discontinuity and shock wave) is that one of the classical "Sod test-case" [95]. Usually (see *e.g.* [98] and [99]), the data are chosen for the Sod test-case so as to get a "sonic rarefaction" (*i.e.* a rarefaction for which $\|\mathbf{u}\| = a$ along a certain characteristic line, see Section 2.4.2), since this wave is a representative benchmark for evaluating the entropic behaviour of a considered numerical scheme (see *e.g.* Note 3.3.12 in Section 3.3.1). Nevertheless, no sonic rarefactions are present in the solution of the considered benchmarks, since a first target of the present work is the simulation of non-cavitating, nearly-incompressible, liquid flows (see Section 1.6) in which sonic conditions can not take place. Conversely, it is of interest here to investigate the behaviour of numerical schemes dealing with low Mach number flows (*e.g.* $\tilde{M} = O(10^{-3}) \div O(10^{-2})$), like those considered in the aforementioned benchmarks. However, it must be remarked that the application of the proposed numerical techniques is by no means restricted to low Mach number flows.

Initial and boundary conditions

The initial discontinuity of the considered RP is located at $x = 0$. Moreover, the space discretization is built in such a way that the right boundary of the cell $C_{\bar{s}}$ ($\bar{s} \in \mathcal{I}, \bar{s} < N_c$) is systematically located at $x = 0$. Hence the following IC is directly derived from that one of the considered RP:

$$\mathbf{q}_i^{(A)0} := \begin{cases} \mathbf{q}_L^{(A)} & i = 1, \dots, \bar{s} \\ \mathbf{q}_R^{(A)} & i = \bar{s}+1, \dots, N_c. \end{cases} \tag{3.27}$$

As far as the BCs are concerned, transmissive conditions are chosen (see *e.g.* [98]), obtained by defining the fictitious state vectors $\mathbf{q}_0^{(A)n}$ and $\mathbf{q}_{N_c+1}^{(A)n}$ (introduced in the relevant paragraph of Section 3.1.1) as follows:

$$\mathbf{q}_0^{(A)n} = \mathbf{q}_1^{(A)n} \ , \ \mathbf{q}_{N_c+1}^{(A)n} = \mathbf{q}_{N_c}^{(A)n} \ , \ n = 0, 1, 2, \dots \tag{3.28}$$

Test-cases

For the sake of simplicity, a uniform space discretization as well as a constant time-step is adopted in (3.10), namely:

$$\mu_i = \mu \ , \ i \in \mathcal{I}$$
$$\delta^n t = \tau \ , \ n = 0, 1, 2, \dots$$

Then, the following CFL-like stability constraint should be enforced when adopting the basic explicit scheme under consideration:

$$\tau \leq c^{(CFL)} \frac{\mu}{s_{max}^n} \ , \ n = 0, 1, 2, \dots \tag{3.29}$$

where:

- s_{max}^n represents the largest wave speed present throughout the computational domain at time-level n;

- $c^{(CFL)}$ denotes a suitable safety coefficient. A possible choice, originally proposed by Godunov, is $c^{(CFL)} = 0.5$, which prevents any wave interaction from taking place within the generic cell C_i. This choice seems to be a little bit strict and a coefficient $0 < c^{(CFL)} \leq 1.0$ is commonly adopted, by assuming that no wave acceleration occurs as a consequence of wave interaction (see *e.g.* [98]).

For the sake of simplicity, both μ and τ are chosen at the beginning of the simulation and the CFL condition (3.29) is only checked during the simulation, at each time-level (in particular, $s^n_{\max}$ is exactly computed by exploiting the relevant relations introduced in Section 2.5.2). The considered test-cases are summarized in Table 3.2, where n_L and n_R respectively represent the number of cells introduced within the "left" and "right" sub-domains (*i.e.* $n_L = \bar{s}$ and $n_R = N_c - \bar{s}$, with $\bar{s}$ appearing in (3.27)); the corresponding numerical solutions are shown in Figures 3.1-3.8.

Table 3.2. Considered test-cases for the discrete scheme (3.10), based on the numerical flux (3.26).

Test-case	Benchmark	μ	(n_L, n_R)	τ
EG1-1	B1	100	$(2,2) \cdot 10^1$	$5 \cdot 10^{-2}$
EG1-2	B1	10	$(2,2) \cdot 10^2$	$5 \cdot 10^{-3}$
EG1-3	B1	1	$(2,2) \cdot 10^3$	$5 \cdot 10^{-4}$
EG1-4	B1	0.1	$(2,2) \cdot 10^4$	$5 \cdot 10^{-5}$
EG2-1	B2	100	$(2,2) \cdot 10^1$	$5 \cdot 10^{-2}$
EG2-2	B2	10	$(2,2) \cdot 10^2$	$5 \cdot 10^{-3}$
EG2-3	B2	1	$(2,2) \cdot 10^3$	$5 \cdot 10^{-4}$
EG2-4	B2	0.1	$(2,2) \cdot 10^4$	$5 \cdot 10^{-5}$

It should be noticed that the left rarefaction appearing in the aforementioned figures is very steep. Such a behaviour is typical of low Mach number flows (see *e.g.* [49]) and can be justified as follows. For the considered flows, the head of the left rarefaction[8] travels with a speed $u_L - a_L \approx u_L - \tilde{a} \approx -\tilde{a}$ where $\tilde{a}$ is the characteristic sound speed ($\tilde{a} \gg |u|$ since $\tilde{M} \ll 1$). Moreover, according to (2.82) the speed of the right shock is $\sigma = u_R + \rho_R^{-1}\zeta(\rho_\star, \rho_R)$ and, in consideration of

[8] The head of the wave is the extreme of the wave region which is in contact with the unperturbed state while the tail is the extreme of the wave region adjacent to the star region [98].

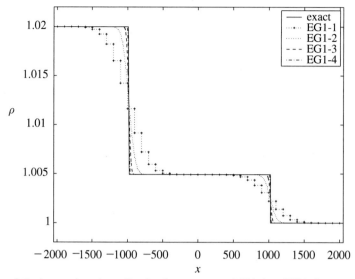

Figure 3.1. Approximation of ρ for the test-cases EG1-1 to EG1-4.

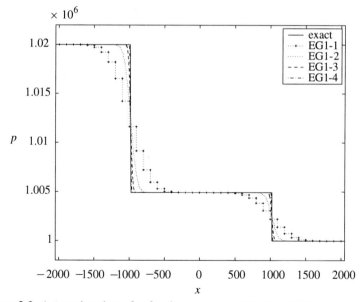

Figure 3.2. Approximation of p for the test-cases EG1-1 to EG1-4.

(2.62), $\sigma > u_R + a_R$. However, since $\rho_\star$ turns out to be close to ρ_R: $\rho_\star/\rho_R = 1 + O(\tilde{M})$, it follows that $\sigma \approx u_R + a_R(1 + O(\tilde{M})) \approx \tilde{a}$ and therefore $\tilde{a}$ can be considered as an acceptable estimate for the shock

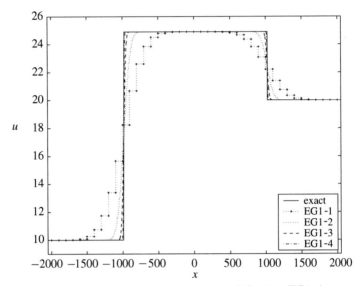

Figure 3.3. Approximation of u for the test-cases EG1-1 to EG1-4.

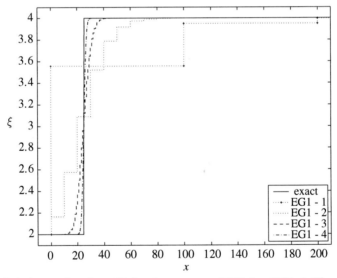

Figure 3.4. Approximation of ξ for the test-cases EG1-1 to EG1-4. The x-range is cut for ease of readability.

speed as well.[9] It is therefore clear that, during a unit time interval ($t_{\text{eval}} = 1$), the flow perturbation extends over an interval having width

---

[9] These considerations incidentally show that, at low Mach numbers, the rarefactions and the shocks approximately travel at the same speed (in absolute value), as confirmed *e.g.* by Figures 3.1-3.3 and Figures 3.5-3.7 (in which the shock and the rarefaction are roughly located at the same distance from the position of the initial discontinuity, *i.e.* $x = 0$).

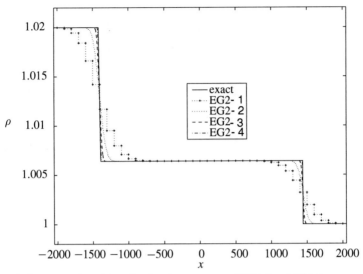

Figure 3.5. Approximation of ρ for the test-cases EG2-1 to EG2-4.

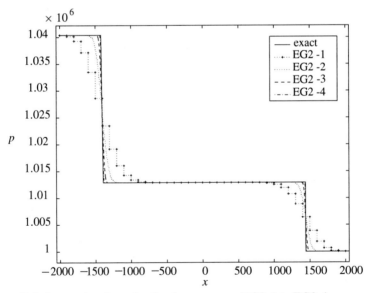

Figure 3.6. Approximation of p for the test-cases EG2-1 to EG2-4.

w_{domain} of the order of $\tilde{a}$:

$$\frac{w_{\text{domain}}}{\tilde{a}} = \text{O}(1) \qquad (3.30)$$

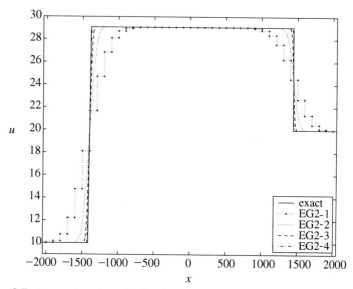

Figure 3.7. Approximation of u for the test-cases EG2-1 to EG2-4.

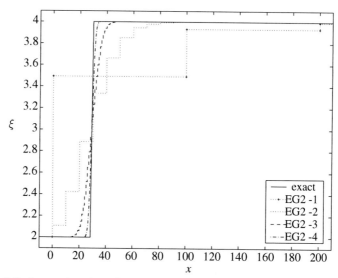

Figure 3.8. Approximation of ξ for the test-cases EG2-1 to EG2-4. The x-range is cut for ease of readability.

As far as the left rarefaction fan is concerned, it is delimited by the following characteristics (compare with (2.47)):

$$\frac{x}{t} = u_L - a_L \;,\quad \frac{x}{t} = u_\star - a_\star$$

and thus, for $t_{\text{eval}} = 1$, its width w_{fan} can be expressed as follows:

$$w_{\text{fan}} = (u_\star - u_L) + (a_L - a_\star)$$

In particular:

- for the benchmark B1, $a_L = a_\star = \tilde{a}$, $u_L = \alpha_L \tilde{M} \tilde{a}$ and $u_\star = \alpha_\star \tilde{M} \tilde{a}$, with α_L and $\alpha_\star$ of the order of the unity and such that $(\alpha_\star - \alpha_L) > 0$. Hence, in the present case w_{fan} reads:

$$w_{\text{fan}} = (\alpha_\star - \alpha_L) \, \tilde{M} \, \tilde{a} \tag{3.31}$$

- for the benchmark B2, $u_L = \gamma_L \tilde{M} \tilde{a}$ and $u_\star = \gamma_\star \tilde{M} \tilde{a}$, with γ_L and $\gamma_\star$ of the order of the unity and such that $(\gamma_\star - \gamma_L) > 0$. Moreover, $\rho_L = \rho_R (1 + \beta_L \tilde{M})$ and $\rho_\star = \rho_R (1 + \beta_\star \tilde{M})$, with β_L and $\beta_\star$ of the order of the unity and such that $(\beta_L - \beta_\star) > 0$. Then, since $a = \sqrt{2\kappa\rho}$ and $\tilde{a} = a_R = \sqrt{2\kappa}$, it follows that $a_L \approx \tilde{a}(1 + \beta_L \tilde{M}/2)$ and $a_\star \approx \tilde{a}(1 + \beta_\star \tilde{M}/2)$. Hence, w_{fan} for the present case reads:

$$w_{\text{fan}} \approx \left[(\gamma_\star - \gamma_L) + \frac{1}{2} (\beta_L - \beta_\star) \right] \tilde{M} \, \tilde{a} . \tag{3.32}$$

In light of (3.31) and (3.32),

$$\frac{w_{\text{fan}}}{\tilde{a}} = \mathrm{O}\left(\tilde{M} \right)$$

for both the considered benchmarks and, by recalling (3.30), it is clear that:

$$\frac{w_{\text{fan}}}{w_{\text{domain}}} = \mathrm{O}\left(\tilde{M} \right)$$

thus motivating the aforementioned observation.

Some entities which can be exploited in order to evaluate the accuracy as well as the computational cost of each simulation are finally reported in Table 3.3, namely:

- an estimate $\tilde{c}^{(\text{CFL})}$ of the CFL coefficient, defined as follows (compare with (3.29)):

$$\tilde{c}^{(\text{CFL})} := \frac{\tau \tilde{s}_{\max}}{\mu} \tag{3.33}$$

where $\tilde{s}_{\max}$ denotes the largest wave speed of the RP associated with the relevant benchmark;

Table 3.3. CFL estimate, CPU time and error estimates for the test-cases reported in Table 3.2.

Test-case	$\tilde{c}^{(\text{CFL})}$	t_{CPU}	$e\,(\rho)$	$e\,(p)$	$e\,(u)$	$e\,(\xi)$
EG1-1	0.51	≈ 0.1 sec.	0.1792	0.1792	8.5733	4.3568
EG1-2	0.51	≈ 1 sec.	0.0967	0.0967	4.6240	2.0348
EG1-3	0.51	≈ 35 sec.	0.0492	0.0492	2.3530	1.0299
EG1-4	0.51	≈ 35 min.	0.0211	0.0211	1.0110	0.4740
EG2-1	0.72	≈ 0.1 sec.	0.1587	0.3185	8.6763	4.5884
EG2-2	0.72	≈ 1 sec.	0.0837	0.1679	4.5757	2.0057
EG2-3	0.72	≈ 35 sec.	0.0392	0.0786	2.1438	1.0515
EG2-4	0.72	≈ 35 min.	0.0141	0.0282	0.7703	0.4493

- the CPU time t_{CPU}, as required on a laptop having the following characteristics: Intel P4 CPU 2.66GHz, 512kB L2 cache, 512MB RAM;
- some error estimates concerning the numerical solution for ρ, p, u and ξ, whose definition is discussed below by considering the generic entity ψ. Let ψ_k^{bench} denote a sequence obtained by sampling the exact solution ψ of the relevant RP in correspondence of the sequence x_k^{bench}, with $k = 1, 2, \ldots, N_{\text{bench}}$, which is fine enough to reproduce the variation of the considered solution almost exactly. Moreover, let ψ_j^{num} denote a discrete approximation of ψ which is constant within each interval $\left(x_j^{\text{num}}, x_{j+1}^{\text{num}}\right)$, with $j = 1, 2, \ldots, (N_{\text{num}} - 1)$. Finally, let D_x represent the following interval:

$$D_x := (-n_L\,\mu,\, n_R\,\mu)$$

Clearly, it is always possible to define the aforementioned sequences over D_x by adjusting the parameters controlling the space discretization in such a way that:

$$\left(x_1^{\text{bench}}, x_{N_{\text{bench}}}^{\text{bench}}\right) = \left(x_1^{\text{num}}, x_{N_{\text{num}}}^{\text{num}}\right) = D_x\,.$$

Furthermore, it is possible to merge x_k^{bench} and x_j^{num} into a new sequence of abscissae, say x_h^{merge} (with $h = 1, 2, \ldots, N_{\text{merge}}$) and to define a linear interpolation of ψ_k^{bench} and ψ_j^{num}, respectively denoted by $\hat{\psi}_h^{\text{bench}}$ and $\hat{\psi}_h^{\text{num}}$, over the new sequence.[10] Then, the following definition can be introduced in order to estimate the error $e\,(\psi)$ connected with the numerical approximation of ψ:

$$
e\,(\psi) := \left(\frac{\displaystyle\int_{D_x} \left(\hat{\psi}_h^{\text{num}} - \hat{\psi}_h^{\text{bench}} \right)^2 \mathrm{d}x}{\displaystyle\int_{D_x} \left(\hat{\psi}_h^{\text{bench}} \right)^2 \mathrm{d}x} \right)^{\frac{1}{2}} \cdot 10^2 \tag{3.34}
$$

where the integration can be carried out by means of the trapezoidal rule [79], consistently with the chosen linear interpolation.
According to Table 3.3, the discrete solution correctly converges towards the exact one.[11] Moreover, the convergence is sub-linear and roughly exhibits the same trend for all the considered entities, as shown in Figures 3.9 and 3.10.

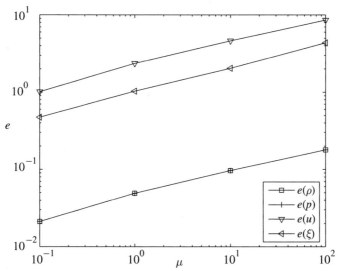

Figure 3.9. Plot of the error estimates for the test-cases EG1-1 to EG1-4 reported in Table 3.3.

---

[10] The linear interpolation is consistent with the piece-wise constant numerical discretization. Moreover, as far as the exact solution is concerned, it should introduce a negligible error in view of the adopted fine sampling.

[11] For the test-cases EG1-1 to EG1-4, $e\,(\rho) = e\,(p)$ by virtue of the direct proportionality between ρ and p which is introduced by the state law associated with the benchmark B1.

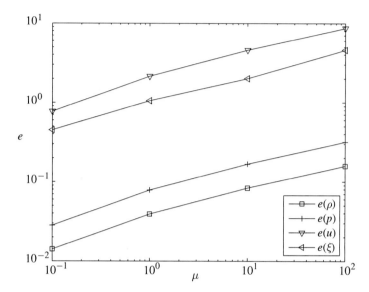

Figure 3.10. Plot of the error estimates for the test-cases EG2-1 to EG2-4 reported in Table 3.3.

3.3. A Roe scheme for generic barotropic state laws

In Section 3.3.1 a Roe numerical flux function applicable to generic barotropic state laws is proposed. In Section 3.3.2 the scheme (3.10) exploiting the proposed numerical flux is validated against an exact solution.

3.3.1. Roe numerical flux

A common numerical flux suitable for incorporation into the Godunov approach (see Section 3.1.1) is that one originally proposed by Roe [84]. According to this method, an approximate RP is suitably introduced at each cell interface and the numerical flux is defined by considering the flux -as obtained by exactly solving the approximate RP- which crosses the interface. In the Roe method, the approximation of the inter-cell flux is obtained "directly" (contrarily to other methods, generally referred to as "approximate-state Riemann solvers", which define the numerical flux by evaluating the analytical one $\mathbf{f}^{(A)}$ in correspondence of a suitably defined state vector); further details can be found in a number of textbooks, *e.g.* [39] and [98] amongst many others.

Definition of the numerical flux $\phi_{LR}^{(A)\text{ROE}}$

The non-linear p.d.e. in (2.52) is locally replaced with the following linear one:

$$\partial_t \mathbf{q}^{(A)} + \tilde{\mathbf{J}}_{LR}^{(A)} \cdot \partial_x \mathbf{q}^{(A)} = 0 \qquad (3.35)$$

where $\tilde{\mathbf{J}}_{LR}^{(A)}$ represents a suitable matrix, called "Roe matrix", depending on the "left" and "right" states $\mathbf{q}_L^{(A)}$ and $\mathbf{q}_R^{(A)}$:

$$\tilde{\mathbf{J}}_{LR}^{(A)} := \tilde{\mathbf{J}}^{(A)}\left(\mathbf{q}_L^{(A)}, \mathbf{q}_R^{(A)}\right). \qquad (3.36)$$

The Roe matrix must verify the following conditions for any couple $(\mathbf{q}_L^{(A)}, \mathbf{q}_R^{(A)})$:

(RM1) $\tilde{\mathbf{J}}_{LR}^{(A)}$ is diagonalizable with real eigenvalues;

(RM2) $\tilde{\mathbf{J}}^{(A)}\left(\mathbf{q}_L^{(A)} \to \mathbf{q}^{(A)\star}, \mathbf{q}_R^{(A)} \to \mathbf{q}^{(A)\star}\right) \to \mathbf{J}^{(A)}\left(\mathbf{q}^{(A)\star}\right)$

  where $\mathbf{J}^{(A)}$ denotes the Jacobian of the original (non linear) flux $\mathbf{f}^{(A)}$ defined in (2.32);

(RM3) let $\Delta^{LR} \mathbf{z}$ denote the variation of the generic vector $\mathbf{z}$ between a "left" state $\mathbf{z}_L$ and a "right" state $\mathbf{z}_R$:

$$\Delta^{LR} \mathbf{z} := \mathbf{z}_R - \mathbf{z}_L. \qquad (3.37)$$

  Then:

$$\Delta^{LR} \mathbf{f}^{(A)} = \tilde{\mathbf{J}}_{LR}^{(A)} \cdot \Delta^{LR} \mathbf{q}^{(A)} \qquad (3.38)$$

  where, of course, $\mathbf{f}_s^{(A)}$ is understood as $\mathbf{f}^{(A)}\left(\mathbf{q}_s^{(A)}\right)$, $s \in \{L, R\}$.

Note 3.3.1. The condition (RM1) ensures that the hyperbolicity of the 1D problem at hand (see Section 2.3.2) is preserved when replacing the original RP with the approximate, linearized, one. The condition (RM2) enforces a natural consistency requirement. The condition (RM3), instead, is imposed by the fact that a unique value for the flux crossing the interface must be obtained by only considering either the left or the right portion of the solution to the considered linearized RP (of course, due to conservation) [98].

By recalling the solution of the considered linearized RP (see *e.g.* [98]), it is possible to express the numerical flux from $\mathbf{q}_L^{(A)}$ to $\mathbf{q}_R^{(A)}$ (along the direction associated with the versor $\hat{e}$ defined in Section 3.1.1) as follows:

$$\phi_{LR}^{(A)\text{ROE}} := \phi_{c,LR}^{(A)\text{ROE}} + \phi_{u,LR}^{(A)\text{ROE}} \qquad (3.39)$$

where:

$$\phi_{c,LR}^{(A)\text{ROE}} := \frac{1}{2} \left(\mathbf{f}_L^{(A)} + \mathbf{f}_R^{(A)} \right) \tag{3.40}$$

$$\phi_{u,LR}^{(A)\text{ROE}} := \mathbf{D}_{LR}^{(A)} \cdot \Delta^{LR} \mathbf{q}^{(A)} \tag{3.41}$$

$$\mathbf{D}_{LR}^{(A)} := -\frac{1}{2} \left| \tilde{\mathbf{J}}_{LR}^{(A)} \right| . \tag{3.42}$$

As far as the definition (3.42) is concerned, the operator $|\cdot|$, defined in (2.1), can be rightfully applied in consideration of the condition (RM1) above.

Note 3.3.2. The expression (3.41) takes into account the wave structure of the linearized RP (in particular, the sudden variation of the solution across the waves propagating along the characteristics [98]) and it is consequently referred to as the "upwind" component of the numerical flux function. The expression (3.40), instead, is often referred to as the "centred" component of the numerical flux function, due to its symmetrical form.

By exploiting the definition (2.2), together with the property (RM3) above, it is possible to recast the numerical flux $\phi_{LR}^{(A)\text{ROE}}$ as follows:

$$\begin{cases} \phi_{LR}^{(A)\text{ROE}} = \mathbf{f}_L^{(A)} + \left(\tilde{\mathbf{J}}_{LR}^{(A)} \right)^- \cdot \Delta^{LR} \mathbf{q}^{(A)} \\ \\ \phi_{LR}^{(A)\text{ROE}} = \mathbf{f}_R^{(A)} - \left(\tilde{\mathbf{J}}_{LR}^{(A)} \right)^+ \cdot \Delta^{LR} \mathbf{q}^{(A)} . \end{cases} \tag{3.43}$$

A Roe matrix for generic barotropic state laws

Clearly, the Roe matrix depends in general on the specific problem under consideration and, in particular, on the specific state law. In the original paper [84], for instance, a Roe matrix is defined for the Euler equations associated with a perfect gas state law.[12] A crucial constraint on the definition of the Roe matrix is given by the condition (RM3) above. In [84], the fulfilment of this condition is obtained by identifying a suitable vector, called "parameter vector", such that both the state vector and the (analytical) flux are homogeneous quadratic functions of it. In the present (barotropic) case, the approach under discussion would encourage to seek a certain vector $\mathbf{z}$ such that $\mathbf{q}^{(A)}$ and $\mathbf{f}^{(A)}$ are homogeneous

---

[12] Different extensions to more complex cases have been proposed in the literature (see *e.g.* [37, 111, 1] and [41]) amongst many others.

quadratic functions of $\mathbf{z}$. Clearly, it is not possible to define such a vector due to the assumed generality of the barotropic curve $p = p(\rho)$. Nevertheless, it is possible to take advantage of the basic idea underlying the considered approach, as described below.

Once the flux $\mathbf{f}^{(A)}$ has been split as follows:

$$\mathbf{f}^{(A)} = \mathbf{f}^{(A)}_H + \mathbf{f}^{(A)}_{NH} \tag{3.44}$$

with:

$$\mathbf{f}^{(A)}_H := \begin{pmatrix} \rho u \\ \rho u^2 \\ \rho u \xi \end{pmatrix}, \quad \mathbf{f}^{(A)}_{NH} := \begin{pmatrix} 0 \\ p \\ 0 \end{pmatrix} \tag{3.45}$$

the variation $\Delta^{LR}\mathbf{f}^{(A)}$ clearly reads:

$$\Delta^{LR}\mathbf{f}^{(A)} = \Delta^{LR}\mathbf{f}^{(A)}_H + \Delta^{LR}\mathbf{f}^{(A)}_{NH}. \tag{3.46}$$

Moreover, once introduced the following vector:

$$\mathbf{z} = \begin{pmatrix} z_1 \\ z_2 \\ z_3 \end{pmatrix} := \begin{pmatrix} \sqrt{\rho} \\ \sqrt{\rho}\, u \\ \sqrt{\rho}\, \xi \end{pmatrix}$$

it is clear that both $\mathbf{q}^{(A)}$ and $\mathbf{f}^{(A)}_H$ are homogeneous quadratic function of $\mathbf{z}$, since:

$$\mathbf{q}^{(A)}(\mathbf{z}) = \begin{pmatrix} z_1^2 \\ z_1 z_2 \\ z_1 z_3 \end{pmatrix}, \quad \mathbf{f}^{(A)}_H(\mathbf{z}) = \begin{pmatrix} z_1 z_2 \\ z_2^2 \\ z_2 z_3 \end{pmatrix}$$

and therefore the following relations hold (by a well-known property of the homogeneous quadratic functions):

$$\Delta^{LR}\mathbf{q}^{(A)} = \mathbf{Q}_{LR} \cdot \Delta^{LR}\mathbf{z}, \quad \Delta^{LR}\mathbf{f}^{(A)}_H = \mathbf{F}_{H,LR} \cdot \Delta^{LR}\mathbf{z} \tag{3.47}$$

where:

$$\mathbf{Q}_{LR} := \partial_{\mathbf{z}}\mathbf{q}^{(A)}\left(\mathbf{z} = \frac{\mathbf{z}_L + \mathbf{z}_R}{2}\right), \quad \mathbf{F}_{H,LR} := \partial_{\mathbf{z}}\mathbf{f}^{(A)}_H\left(\mathbf{z} = \frac{\mathbf{z}_L + \mathbf{z}_R}{2}\right).$$

Then, by combining the equations in (3.47), the following relation is immediately obtained:

$$\Delta^{LR}\mathbf{f}^{(A)}_H = \hat{\mathbf{J}}^{(A)}_{LR} \cdot \Delta^{LR}\mathbf{q}^{(A)}$$

with:

$$\hat{\mathbf{J}}^{(A)}_{LR} := \mathbf{F}_{H,LR} \cdot \mathbf{Q}_{LR}^{-1}$$

and (3.46) can be finally recast as follows:

$$\Delta^{LR}\mathbf{f}^{(A)} = \hat{\mathbf{J}}_{LR}^{(A)} \cdot \Delta^{LR}\mathbf{q}^{(A)} + \Delta^{LR}\mathbf{f}_{NH}^{(A)} \,. \tag{3.48}$$

Straightforward computations lead to the following representation, in particular, for $\hat{\mathbf{J}}_{LR}^{(A)}$:

$$\hat{\mathbf{J}}_{LR}^{(A)} = \begin{pmatrix} 0 & 1 & 0 \\ -u_{LR}^2 & 2u_{LR} & 0 \\ -u_{LR}\,\xi_{LR} & \xi_{LR} & u_{LR} \end{pmatrix} \tag{3.49}$$

where (subscripts "L" and "R", as applied to vector components, are understood in the sequel):

$$\begin{cases} u_{LR} := \dfrac{\sqrt{\rho_L}\,u_L + \sqrt{\rho_R}\,u_R}{\sqrt{\rho_L} + \sqrt{\rho_R}} \\[4mm] \xi_{LR} := \dfrac{\sqrt{\rho_L}\,\xi_L + \sqrt{\rho_R}\,\xi_R}{\sqrt{\rho_L} + \sqrt{\rho_R}} \,. \end{cases} \tag{3.50}$$

By starting from (3.48), it is possible to to match the condition (3.38) and, consequently, to define a Roe matrix for the barotropic case under consideration, as shown in the following:

Proposition 3.3.3. *A Roe matrix $\tilde{\mathbf{J}}_{LR}^{(A)}$ applicable when considering a generic barotropic state law reads:*

$$\tilde{\mathbf{J}}_{LR}^{(A)} = \begin{pmatrix} 0 & 1 & 0 \\ a_{LR}^2 - u_{LR}^2 & 2u_{LR} & 0 \\ -u_{LR}\,\xi_{LR} & \xi_{LR} & u_{LR} \end{pmatrix} \tag{3.51}$$

where:

$$a_{LR} := \begin{cases} \left(\dfrac{\Delta^{LR}p}{\Delta^{LR}\rho}\right)^{\frac{1}{2}} & \text{if } \rho_R \neq \rho_L \\[4mm] a(\rho_\star) & \text{if } \rho_R = \rho_L = \rho_\star \end{cases} \tag{3.52}$$

Proof. At a first step, a matrix $\check{\mathbf{J}}_{LR}^{(A)}$ is sought such that:

$$\begin{pmatrix} 0 \\ \Delta^{LR}p \\ 0 \end{pmatrix} = \Delta^{LR}\mathbf{f}_{NH}^{(A)} = \check{\mathbf{J}}_{LR}^{(A)} \cdot \Delta^{LR}\mathbf{q}^{(A)} \tag{3.53}$$

Let α_{mn} ($m, n \in \{1, 2, 3\}$) denote the mn-th component of $\check{\mathbf{J}}_{LR}^{(A)}$. Then, $\alpha_{1n} = \alpha_{3n} = 0$ due to the mutual independence of the state vector components (*i.e.* ρ, ρu and $\rho\xi$), while $\alpha_{22} = \alpha_{23} = 0$ by virtue of the barotropic state law (1.3). Hence, (3.53) reduces to the following scalar equation:

$$\Delta^{LR} p = \alpha_{21} \Delta^{LR} \rho . \tag{3.54}$$

When $\rho_R = \rho_L$ the above equation is trivially verified regardless of the specific value of α_{21} while, for $\rho_R \neq \rho_L$ it necessarily follows that:

$$\alpha_{21} = \frac{\Delta^{LR} p}{\Delta^{LR} \rho} , \quad \rho_R \neq \rho_L$$

where the divided difference is positive, due to the strict monotonicity of $p(\rho)$ assumed in (1.4). Hence, by choosing $\alpha_{21} = a_{LR}^2$ with a_{LR} defined in (3.52), a continuous (*i.e.* prolongated by continuity) solution is obtained. As a result, the expression of $\check{\mathbf{J}}_{LR}^{(A)}$ reads:

$$\check{\mathbf{J}}_{LR}^{(A)} = \begin{pmatrix} 0 & 0 & 0 \\ a_{LR}^2 & 0 & 0 \\ 0 & 0 & 0 \end{pmatrix}. \tag{3.55}$$

By substituting (3.53) into (3.48) it is evident that the following matrix:

$$\tilde{\mathbf{J}}_{LR}^{(A)} = \hat{\mathbf{J}}_{LR}^{(A)} + \check{\mathbf{J}}_{LR}^{(A)} \tag{3.56}$$

satisfies the condition (3.38) (*i.e.* the condition (RM3) above). Furthermore, it is straightforward to verify that the considered matrix $\tilde{\mathbf{J}}_{LR}^{(A)}$ also satisfies the aforementioned conditions (RM1) and (RM2) and, therefore, it is a suitable Roe matrix for the generic barotropic case at hand. As far as its representation is concerned, by substituting (3.49) and (3.55) into (3.56), the expression (3.51) is immediately obtained. This completes the proof. $\square$

Note 3.3.4. While u_{LR} and ξ_{LR} in (3.50) are well-known "Roe averages" [84], a_{LR} in (3.52) represents an average value (hereafter referred to as Roe average as well) which is specific to the present (generic) barotropic case. For instance, it can be also exploited when considering the well-known homogeneous shallow water equations, since they can be derived from the considered conservation laws (see Note 2.5.1 in Section 2.5.1). Indeed, the expression (3.52) generalizes the relevant one defined in [99] for the shallow water case.

Note 3.3.5. It may be worth mentioning that, as far as its numerical implementation is concerned, a_{LR} should be defined as follows:

$$a_{LR} := \begin{cases} \left(\dfrac{\Delta^{LR} p}{\Delta^{LR} \rho} \right)^{1/2} & \text{if } \mid \rho_R - \rho_L \mid > \epsilon_\rho \\ a\left(\rho = \varrho\left(\rho_L, \rho_R\right) \right) & \text{if } \mid \rho_R - \rho_L \mid < \epsilon_\rho \end{cases}$$

where ϵ_ρ is a suitable numerical threshold and $\varrho\left(\rho_L, \rho_R\right)$ is an average value (e.g. a geometrical mean) such that $\varrho\left(\rho_L \to \rho^\star, \rho_R \to \rho^\star\right) \to \rho^\star$.

Note 3.3.6. The expression of the Jacobian $\mathbf{J}^{(A)}$ defined in (2.32) reads:

$$\mathbf{J}^{(A)} = \begin{pmatrix} 0 & 1 & 0 \\ a^2 - u^2 & 2u & 0 \\ -u\,\xi & \xi & u \end{pmatrix}. \tag{3.57}$$

Once noticed the formal similarity between (3.57) and (3.51) and by interpreting $\mathbf{J}^{(A)}$ in (3.57) as a function of a, u and ξ, the following relation clearly holds:[13]

$$\tilde{\mathbf{J}}^{(A)}_{LR} = \mathbf{J}^{(A)}\left(a = a_{LR}, u = u_{LR}, \xi = \xi_{LR} \right). \tag{3.58}$$

Note 3.3.7. The Roe matrix (3.51) keeps the same representation even when the "left" and "right" states are interchanged, due to the "symmetrical" definition of its components, namely (3.50) and (3.52).

Note 3.3.8. By neglecting the third row and the third column of (3.51), a Roe matrix for the basic-1D equations (see Section 2.2.3) is obtained, which has been previously introduced in [38].

Note 3.3.9. In [91], a Roe matrix for the basic-1D equations (see Section 2.2.3) completed with the energy balance is defined. The averages appearing in [91] can be derived from those obtained in [37] for the case of a generic state law of the form $p = p(\rho, e_i)$, e_i denoting the internal energy per unit mass.

Note 3.3.10. It is straightforward to extend the Roe matrix (3.51) to the case of $m > 1$ passive scalars. When adopting, for instance, the definition

---

[13] An analogous relation holds for the Euler equations associated with a perfect gas state law [84].

(2.23) for the state vector $\mathbf{q}^{(A)}$ ($m = 2$), the considered Roe matrix reads:

$$
\tilde{\mathbf{J}}_{LR}^{(A)} =
\begin{pmatrix}
0 & 1 & 0 & 0 \\
a_{LR}^2 - u_{LR}^2 & 2u_{LR} & 0 & 0 \\
-u_{LR}\,\xi_{LR} & \xi_{LR} & u_{LR} & 0 \\
-u_{LR}\,\eta_{LR} & \eta_{LR} & 0 & u_{LR}
\end{pmatrix}
\tag{3.59}
$$

with η_{LR} defined analogously to ξ_{LR} in (3.50).

Definition of the numerical flux $\phi_{ij}^{(A)\mathrm{ROE}}$

Let $\mathbf{q}_i^{(A)}$, $\mathbf{q}_j^{(A)}$ and $\hat{\mathbf{v}}_{ij}$, with $j \in \pi_i$, be defined as in Section 3.1.1. In particular, $\mathbf{q}_i^{(A)}$ and $\mathbf{q}_j^{(A)}$ can respectively represent either a couple of "left" and "right" states or vice-versa. It is possible to exploit the numerical flux (3.39)-(3.42) in order to define a Roe numerical flux from $\mathbf{q}_i^{(A)}$ to $\mathbf{q}_j^{(A)}$, as described below.

As far as the Roe matrix is concerned, in view of the "symmetry" already pointed out in Note 3.3.7 above, it is straightforward to generalize (3.51) as follows:

$$
\tilde{\mathbf{J}}_{ij}^{(A)} :=
\begin{pmatrix}
0 & 1 & 0 \\
a_{ij}^2 - u_{ij}^2 & 2u_{ij} & 0 \\
-u_{ij}\,\xi_{ij} & \xi_{ij} & u_{ij}
\end{pmatrix}
\tag{3.60}
$$

with:

$$
\begin{cases}
u_{ij} := \dfrac{\sqrt{\rho_i}\, u_i + \sqrt{\rho_j}\, u_j}{\sqrt{\rho_i} + \sqrt{\rho_j}} \\[3mm]
\xi_{ij} := \dfrac{\sqrt{\rho_i}\, \xi_i + \sqrt{\rho_j}\, \xi_j}{\sqrt{\rho_i} + \sqrt{\rho_j}}
\end{cases}
\tag{3.61}
$$

$$
a_{ij} :=
\begin{cases}
\left(\dfrac{\Delta^{ij} p}{\Delta^{ij} \rho}\right)^{\frac{1}{2}} & \text{if } \rho_j \neq \rho_i \\[3mm]
a(\rho_\star) & \text{if } \rho_j = \rho_i = \rho_\star
\end{cases}
\tag{3.62}
$$

and, of course:

$$
\Delta^{ij} (\cdot) := (\cdot)_j - (\cdot)_i .
\tag{3.63}
$$

Then, once recalled the definition of s_{ij} given in (3.6), it is possible to define the Roe numerical under consideration as follows:

$$\phi^{(A)\text{ROE}} \left(\mathbf{q}_i^{(A)}, \mathbf{q}_j^{(A)}, \hat{\mathbf{v}}_{ij} \right) := \phi_{ij}^{(A)\text{ROE}} \ , \ j \in \pi_i \qquad (3.64)$$

with:

$$\phi_{ij}^{(A)\text{ROE}} \ := \ \phi_{c,ij}^{(A)\text{ROE}} + \phi_{u,ij}^{(A)\text{ROE}} \qquad (3.65)$$

$$\phi_{c,ij}^{(A)\text{ROE}} \ := \ \frac{1}{2} s_{ij} \left(\mathbf{f}_i^{(A)} + \mathbf{f}_j^{(A)} \right) \qquad (3.66)$$

$$\phi_{u,ij}^{(A)\text{ROE}} \ := \ \mathbf{D}_{ij}^{(A)} \cdot \Delta^{ij} \mathbf{q}^{(A)} \qquad (3.67)$$

$$\mathbf{D}_{ij}^{(A)} \ := \ -\frac{1}{2} \left| s_{ij} \tilde{\mathbf{J}}_{ij}^{(A)} \right| \qquad (3.68)$$

where, of course, $\mathbf{f}_i^{(A)}$ is understood as $\mathbf{f}^{(A)}(\mathbf{q}_i^{(A)})$.

The expressions (3.66)-(3.68) are defined by considering the following RP which generalizes (2.52):

$$\begin{cases} \partial_t \mathbf{q}^{(A)} + \partial_x \left(s_{ij} \, \mathbf{f}^{(A)} \right) = 0 \quad \text{in} \quad \mathbb{R} \times (0, \infty) \\ \\ \mathbf{q}^{(A)} = \begin{cases} \mathbf{q}_i^{(A)} & \text{if } x < 0 \\ \mathbf{q}_j^{(A)} & \text{if } x > 0 \end{cases} \quad \text{on} \quad \mathbb{R} \times \{t = 0\} \end{cases} \qquad (3.69)$$

with $i \in \mathcal{I}$ and $j \in \pi_i$. Indeed:

- for $j = i + 1$ (3.69) reduces to (2.52), with $\mathbf{q}_L^{(A)} = \mathbf{q}_i^{(A)}$ and $\mathbf{q}_R^{(A)} = \mathbf{q}_j^{(A)}$;
- also for $j = i - 1$ (3.69) reduces to (2.52), at the only cost of reversing the orientation of the x-axis (accordingly to $\hat{\mathbf{v}}_{ij} = -\hat{e}$); in this case, $\mathbf{q}_L^{(A)} = \mathbf{q}_j^{(A)}$ and $\mathbf{q}_R^{(A)} = \mathbf{q}_i^{(A)}$.

In both cases, the generalized analytical flux $s_{ij} \, \mathbf{f}^{(A)}$ must be considered when defining the Roe linearization and the corresponding Roe matrix is clearly given by $s_{ij} \tilde{\mathbf{J}}_{ij}^{(A)}$, with $\tilde{\mathbf{J}}_{ij}^{(A)}$ defined in (3.60).

Note 3.3.11. It should be noticed that the Roe numerical flux (3.64) trivially satisfies the consistency property (3.8). The conservation property (3.7) is satisfied as well. Indeed, while $s_{ji} = -s_{ij}$ and $\tilde{\mathbf{J}}_{ji}^{(A)} = \tilde{\mathbf{J}}_{ij}^{(A)}$, the following relation clearly holds, due to the definition of the operator $| \cdot |$ introduced in (2.1):

$$\left| s_{ji}\, \tilde{\mathbf{J}}_{ji}^{(A)} \right| = \left| s_{ij}\, \tilde{\mathbf{J}}_{ij}^{(A)} \right| . \tag{3.70}$$

In consideration of the equality (3.70), the introduction of s_{ij} into (3.68) may appear not necessary. Nevertheless, for the sake of consistency with (3.42), the operator $| \cdot |$ should be applied to the Roe matrix of the considered, namely $s_{ij}\, \tilde{\mathbf{J}}_{ij}^{(A)}$. Moreover, it is compulsory to work with the proper Roe matrix $s_{ij}\, \tilde{\mathbf{J}}_{ij}^{(A)}$ if alternative formulations of the type of (3.43) are sought (as, for instance, in Section 3.5.1), since:

$$\left(s_{ij}\, \tilde{\mathbf{J}}_{ij}^{(A)} \right)^{\pm} \neq \left(\tilde{\mathbf{J}}_{ij}^{(A)} \right)^{\pm} .$$

In particular, the following relations hold:

$$\left(s_{ji}\, \tilde{\mathbf{J}}_{ji}^{(A)} \right)^{\pm} = - \left(s_{ij}\, \tilde{\mathbf{J}}_{ij}^{(A)} \right)^{\mp} \tag{3.71}$$

where the definition of the operators $(\cdot)^{+}$ and $(\cdot)^{-}$ is introduced in (2.2).

Note 3.3.12. It is known from the literature (see *e.g.* [39, 84] and [98]) that non-physical results may arise when exploiting the Roe scheme (3.64)-(3.68), due to the fact that the solution to the linearized problem, always consisting of discontinuities (see *e.g.* [98]), does not provide a correct approximation of continuous waves, like rarefactions. In a practical computational set up, however, problems generally arise when dealing with sonic rarefactions (*i.e.* rarefactions for which $\|\mathbf{u}\| = a$ along a certain characteristic line, see Section 2.4.2): these show up in the form of discontinuities violating the RH condition (2.36). Besides the classical correction technique introduced in [43], various "entropy fixes" have been proposed in the literature to counteract this problem (see [39, 64] and [98] for a comprehensive list of references).

It must be stated in advance that, despite the importance of the issue under consideration, no entropy fixes are considered in the present document, to be applied to the proposed Roe numerical flux (3.64)-(3.68). Indeed, the time-schedule of the industrial project this work was based on, imposed to directly concentrate on the simulation of non-cavitating flows, as a compulsory intermediate step towards the simulation of cavitation, as mentioned in Section 1.6. Then, in consideration of the fact that pure liquid flows are nearly-incompressible (and therefore far from allowing for sonic conditions to take place), the investigation of the entropic behaviour of the considered numerical schemes was initially postponed and the numerical method developed for non-cavitating flows has been exploited for cavitating simulations as well (see Chapter 6). On the other hand, when cavitation occurs a transonic regime is systematically encountered (see Section 1.4) and the entropic behaviour of the considered Roe scheme must be assessed; according to the author, a further study should be devoted to this issue.

3.3.2. Numerical results

Benchmarks

The benchmarks introduced and discussed in Section 3.2.2 (see Table 3.1) are considered here for validating the discrete scheme (3.10), based on the proposed numerical flux (3.64)-(3.68).

Initial and boundary conditions

The IC and the BCs introduced in Section 3.2.2, namely (3.27) and (3.28), are adopted here.

Test-cases

In order to directly compare the proposed Roe numerical flux with the Godunov one proposed in Section 3.2.1, the test-cases introduced in Section 3.2.2 (see Table 3.2) are considered here. They are reported in Table 3.4, where the labels in the first column (different from those in Table 3.2) remind that the Roe numerical flux is exploited here. The corresponding numerical solutions are shown in Figures 3.11-3.18.

Table 3.4. Considered test-cases for the discrete scheme (3.10), based on the numerical flux (3.64)-(3.68).

Test-case	Benchmark	μ	(n_L, n_R)	τ
ER1-1	B1	100	$(2, 2) \cdot 10^1$	$5 \cdot 10^{-2}$
ER1-2	B1	10	$(2, 2) \cdot 10^2$	$5 \cdot 10^{-3}$
ER1-3	B1	1	$(2, 2) \cdot 10^3$	$5 \cdot 10^{-4}$
ER1-4	B1	0.1	$(2, 2) \cdot 10^4$	$5 \cdot 10^{-5}$
ER2-1	B2	100	$(2, 2) \cdot 10^1$	$5 \cdot 10^{-2}$
ER2-2	B2	10	$(2, 2) \cdot 10^2$	$5 \cdot 10^{-3}$
ER2-3	B2	1	$(2, 2) \cdot 10^3$	$5 \cdot 10^{-4}$
ER2-4	B2	0.1	$(2, 2) \cdot 10^4$	$5 \cdot 10^{-5}$

As for the case of the Godunov flux, some entities which can be exploited for evaluating the accuracy and the computational cost of the considered

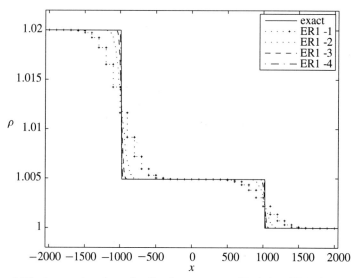

Figure 3.11. Approximation of ρ for the test-cases ER1-1 to ER1-4.

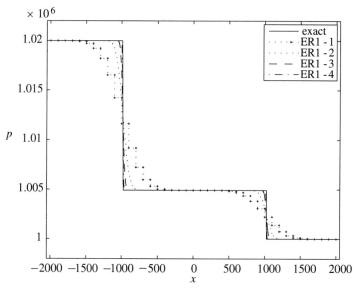

Figure 3.12. Approximation of p for the test-cases ER1-1 to ER1-4.

simulations are reported in Table 3.5 (the definition of the relevant entities is reported in Section 3.2.2, in correspondence of the introduction of Table 3.3). It should be noticed that:

- the column reporting the estimate $\tilde{c}^{(\mathrm{CFL})}$ of the CFL coefficient is clearly identical to the corresponding one in Table 3.3 since all the

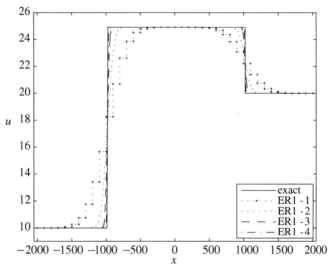

Figure 3.13. Approximation of u for the test-cases ER1-1 to ER1-4.

Table 3.5. CFL estimate, CPU time and error estimates for the test-cases reported in Table 3.4.

Test-case	$\tilde{c}^{(CFL)}$	t_{CPU}	$e(\rho)$	$e(p)$	$e(u)$	$e(\xi)$
ER1-1	0.51	≈ 0.1 sec.	0.1792	0.1792	8.5737	4.3596
ER1-2	0.51	≈ 1 sec.	0.0967	0.0967	4.6240	2.0329
ER1-3	0.51	≈ 35 sec.	0.0492	0.0492	2.3530	1.0297
ER1-4	0.51	≈ 40 min.	0.0211	0.0211	1.0110	0.4740
ER2-1	0.72	≈ 0.1 sec.	0.1587	0.3185	8.6766	4.5910
ER2-2	0.72	≈ 1 sec.	0.0837	0.1679	4.5757	2.0039
ER2-3	0.72	≈ 35 sec.	0.0392	0.0786	2.1438	1.0514
ER2-4	0.72	≈ 40 min.	0.0141	0.0282	0.7703	0.4492

parameters involved in the definition of $\tilde{c}^{(CFL)}$, namely τ, μ and $\tilde{s}_{max}$ in (3.33), have the same value for corresponding test-cases;

• the CPU time (on a laptop with Intel P4 CPU 2.66GHz, 512kB L2 cache, 512MB RAM) is similar to that one required when adopting the Godunov flux. According to the author, the discrepancy between the test-cases ER1-4 and EG1-4 (or ER2-4 and EG2-4) may be due to some differences in the implementation of the considered schemes;[14]

---

[14] The implementation of the Roe scheme, in particular, has been developed by repeatedly calling some BLAS (Basic Linear Algebra Subprograms, see www.netlib.org/blas) routines, even when dealing with very small (*i.e.* 2-4 components) arrays. This point, together with the fact that no

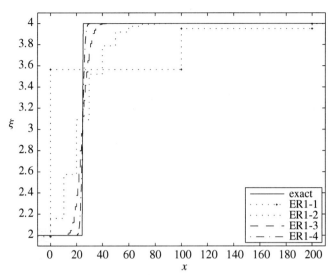

Figure 3.14. Approximation of ξ for the test-cases ER1-1 to ER1-4. The x-range is cut for ease of readability.

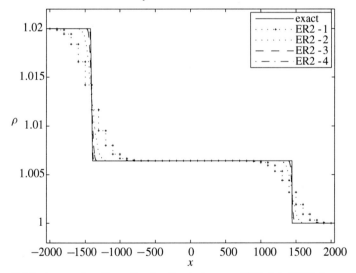

Figure 3.15. Approximation of ρ for the test-cases ER2-1 to ER2-4.

- by comparing the error estimates in Tables 3.3 and 3.5, the discrete scheme based on the Roe flux turns out to behave similarly to that one exploiting the Godunov flux. This result could be partly related to the fact that, for low Mach numbers, the shocks and the rarefac-

---

ad-hoc tuning has been performed for the aforementioned external library (in consideration of the underlying computing platform), may have introduced a certain amount of computational overhead, which becomes more evident for the longest simulations.

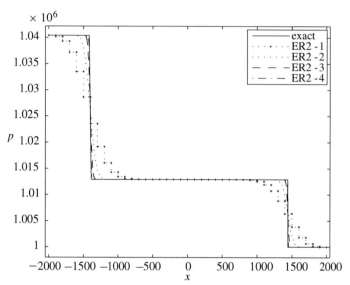

Figure 3.16. Approximation of p for the test-cases ER2-1 to ER2-4.

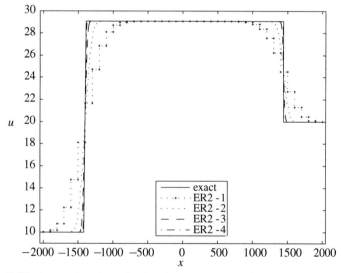

Figure 3.17. Approximation of u for the test-cases ER2-1 to ER2-4.

tions appearing in the solution of the original (non-linear) RP tend to be close to the corresponding discontinuities in the Roe-linearized RP. This observation can be applied locally, at the generic cell interface between the state vectors $\mathbf{q}_i^{(A)}$ and $\mathbf{q}_j^{(A)}$. For the sake of illustration, let $\tilde{a}$ and $\tilde{M} \ll 1$ respectively denote a characteristic sound speed and a characteristic Mach number of the considered flow field.

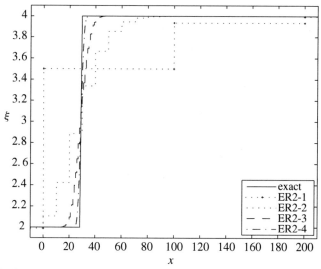

Figure 3.18. Approximation of ξ for the test-cases ER2-1 to ER2-4. The x-range is cut for ease of readability.

Then, the aforementioned discontinuities in the Roe-linearized RP travel with a speed $u_{ij} \pm a_{ij}$. Since for low Mach numbers $a_{ij} \approx \tilde{a}$ and $u_{ij} \leq \max{(u_i, u_j)} \approx \pm \tilde{a}\tilde{M}$, it follows that the speed under consideration can be approximated by $\pm \tilde{a}(1 + \mathrm{O}(\tilde{M}))$, as for the shocks and the rarefactions -which originate, in practice, discontinuities like the shocks- of the non-linear problem (see the relevant paragraph in Section 3.2.2). As far as the contact discontinuity is concerned, its speed is given by u_{ij} for the Roe-linearized problem while for the non-linear one is given by $u_\star$. Let the distance of both ρ_i and ρ_j from a certain reference value, say $\tilde{\rho}$, be of the order of $\tilde{\rho}\tilde{M}$ (consistently with the fact that small density variations take place in nearly-incompressible flows); then, from the definitions (3.61) and (2.84) it follows that for low Mach numbers:

$$u_{ij} \approx \frac{u_i + u_j}{2} \left(1 + \mathrm{O}\left(\tilde{M}\right)\right)$$

$$u_\star \approx \frac{u_i + u_j}{2} \left(1 + \mathrm{O}\left(1\right)\right)$$

and therefore the asymptotic behaviour of the considered speeds is different. Furthermore, while u_{ij} is always contained between $\min(u_i, u_j)$ and $\max{(u_i, u_j)}$ since the relations in (3.61) are convex combinations, $u_\star$ does not necessarily belong to the aforementioned interval, as shown *e.g.* in Figures 3.13 and 3.17. Nevertheless, the difference under consideration may not be accurately perceived by the error estimate $e(\xi)$ defined in (3.34), since the numerator of (3.34) for the present case is small with respect to the denominator (indeed the variation of ξ across the contact discontinuity is abrupt) for both cases.

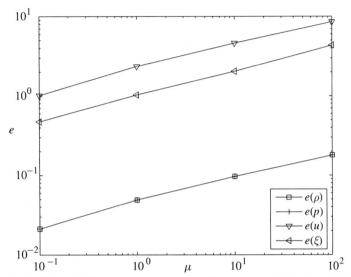

Figure 3.19. Plot of the error estimates for the test-cases ER1-1 to ER1-4 reported in Table 3.5.

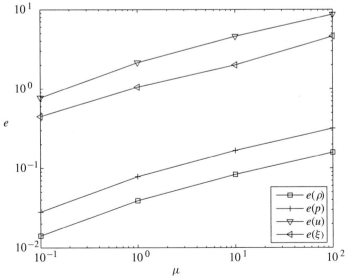

Figure 3.20. Plot of the error estimates for the test-cases ER2-1 to ER2-4 reported in Table 3.5.

According to Table 3.5, the discrete solution correctly converges towards the exact one.[15] Moreover, the convergence is sub-linear and roughly exhibits the same trend for all the considered entities, as shown in Figures 3.19 and 3.20.

3.4. Preconditioning of the Roe scheme for low Mach number flows

It is known from the literature that classical numerical schemes conceived for compressible flows (in particular, a variety of finite volume methods) exhibit accuracy problems when dealing with nearly-incompressible ones. In [42], for instance, the compressible Euler equations coupled with a perfect gas state law are considered. More in detail, the low Mach number asymptotic solution is investigated, as obtained by starting from the continuous formulation as well as from a semi-discrete one, of the type of (3.9) and based on a Roe flux function. It is shown, in particular, that the semi-discrete solution can exhibit pressure variations in space higher than those associated with the analytical one. Furthermore, a suitably modified numerical flux function is consequently introduced in order to counteract this discrepancy.

An investigation of the same type as that one in [42] is performed in the present section, by considering the basic-1D partial differential system (2.18) for the sake of simplicity.[16] More in detail, consistently with the fact that nearly-incompressible flow regions generally do not exhibit discontinuities, smooth (*i.e.* differentiable enough) solutions to (2.18) are considered.[17]

The semi-discrete formulation, based on the Roe numerical flux proposed in Section 3.3.1, which corresponds to (2.18) reads (compare with (3.9)):

$$\mu_i \frac{d}{dt} \mathbf{q}_i^{(x)} + \sum_{j \in \pi_i} \boldsymbol{\phi}^{(x)\text{ROE}} \left(\mathbf{q}_i^{(x)}, \mathbf{q}_j^{(x)}, \hat{\boldsymbol{v}}_{ij} \right) = \mathbf{0} , \; i \in \mathcal{I} \qquad (3.72)$$

---

[15] For the test-cases ER1-1 to ER1-4, $e(\rho) = e(p)$ by virtue of the direct proportionality between ρ and p which is introduced by the state law associated with the benchmark B1.

[16] Indeed, as remarked in Note 2.2.2 (Section 2.2.4), the presence of the passive scalar does not affect the underlying 1D flow field. Hence, in principle, it suffices to consider the mass and momentum balances in order to highlight the problem under consideration.

[17] By virtue of this position, the solution to a RP associated with (2.18) is not of interest in the present case, even when it involves two rarefactions. Indeed, it is not difficult to see that both ρ -and therefore p- and u are strictly monotonic functions of x -for a fixed time- within the rarefaction fans (their derivative being proportional to the convexity marker $c(\rho)$ defined in (2.55)); consequently, the considered solution is continuous (since there is no contact discontinuity in the basic-1D case) but it is not differentiable.

where $\mathbf{q}^{(x)}$ is given by (2.15) and $\boldsymbol{\phi}^{(x)\text{ROE}}(\cdot, \cdot, \cdot)$ is straightforwardly derived from (3.64)-(3.68) as follows:

$$\boldsymbol{\phi}^{(x)\text{ROE}}\left(\mathbf{q}_i^{(x)}, \mathbf{q}_j^{(x)}, \hat{v}_{ij}\right) := \boldsymbol{\phi}_{ij}^{(x)\text{ROE}} \, , \quad j \in \pi_i \tag{3.73}$$

with:

$$\boldsymbol{\phi}_{ij}^{(x)\text{ROE}} := \boldsymbol{\phi}_{c,ij}^{(x)\text{ROE}} + \boldsymbol{\phi}_{u,ij}^{(x)\text{ROE}}$$

$$\boldsymbol{\phi}_{c,ij}^{(x)\text{ROE}} := \frac{1}{2} s_{ij} \left(\mathbf{f}_i^{(x)} + \mathbf{f}_j^{(x)}\right) \tag{3.74}$$

$$\boldsymbol{\phi}_{u,ij}^{(x)\text{ROE}} := \mathbf{D}_{ij}^{(x)} \cdot \Delta^{ij} \mathbf{q}^{(x)}$$

$$\mathbf{D}_{ij}^{(x)} := -\frac{1}{2} \left| s_{ij} \, \tilde{\mathbf{J}}_{ij}^{(x)} \right| \tag{3.75}$$

where s_{ij} is defined in (3.6) and, of course, $\mathbf{f}_i^{(x)}$ is understood as $\mathbf{f}^{(x)}(\mathbf{q}_i^{(x)})$, with $\mathbf{f}^{(x)}(\cdot)$ defined in (2.16). Moreover, the Roe matrix $\tilde{\mathbf{J}}_{ij}^{(x)}$ in (3.75) is defined as follows (see Note 3.3.8 in Section 3.3.1):

$$\tilde{\mathbf{J}}_{ij}^{(x)} := \begin{pmatrix} 0 & 1 \\ a_{ij}^2 - u_{ij}^2 & 2 u_{ij} \end{pmatrix} \tag{3.76}$$

where u_{ij} and a_{ij} are respectively given by (3.61) and (3.62).

By following [42], an asymptotic study is performed in Section 3.4.1; it is shown, in particular, that also in the present (barotropic) case, for nearly-incompressible flows, there is a discrepancy between the behaviour of the solution of the continuous problem (2.18) and that one of the semi-discrete one (3.72). In Section 3.4.2 a concise introduction to preconditioning techniques for low Mach flows is reported. Then, in Section 3.4.3 the preconditioning technique originally proposed in [42] is applied to the proposed numerical flux (3.64)-(3.68) (in particular to its purely 1D counterpart (3.73)-(3.75)), with the aim of eliminating the discrepancy under consideration. Finally, in Section 3.4.4 a discrete scheme based on the proposed preconditioned numerical flux is validated against a nearly-exact solution.

3.4.1. Low Mach number asymptotic study

Non-dimensionalization

In order to determine the behaviour of low Mach number asymptotic solutions, both the continuous problem (2.18) and the semi-discrete one

(3.72) are non-dimensionalized by means of the following reference entities:

$$
\begin{cases}
x_{\text{ref}} \\
u_{\text{ref}} := \max\limits_{x \in D_x} u(x, t = 0) \\
\rho_{\text{ref}} := \max\limits_{x \in D_x} \rho(x, t = 0) \\
a_{\text{ref}} := \max\limits_{x \in D_x} a(x, t = 0) \\
t_{\text{ref}} := x_{\text{ref}} u_{\text{ref}}^{-1} \\
p_{\text{ref}} := \rho_{\text{ref}} a_{\text{ref}}^2
\end{cases}
\tag{3.77}
$$

where x_{ref} denotes a suitable reference length and D_x represents the x-domain (the remaining entities being understood). More in detail, each non-dimensional entity (namely the flow variables, the Roe averages, any relevant function like the state law, etc...) is defined dividing its dimensional counterpart by the proper reference value.

It must be noticed that a reference sound speed a_{ref} is explicitly introduced in (3.77) in order to directly take into account the compressibility effects.[18]

Note 3.4.1. No specific symbols are introduced in order to distinguish between the non-dimensional entities and their dimensional counterparts, for the sake of simplicity.

The non-dimensional form of the continuous system (2.18), which is introduced in Section A.1 for ease of presentation, reads:

$$
\begin{cases}
\partial_t (\rho) = \Psi_c^{(0)} \\
\partial_t (\rho u) = M_\star^{-2} \Theta_c^{(-2)} + \Theta_c^{(0)}
\end{cases}
\tag{3.78}
$$

where the relevant coefficients are defined in the aforementioned section.[19] The non-dimensional parameter $M_\star$ appearing in (3.78) is defined as follows:

$$
M_\star := \frac{u_{\text{ref}}}{a_{\text{ref}}}
\tag{3.79}
$$

and plays a key role in the asymptotic study under consideration. Indeed, the nearly-incompressible limit of the considered equations is obtained for $M_\star \to 0$ (and therefore $M_\star$ is considered as a characteristic Mach number of the flow field).

---

[18] This position is not in contrast with that one mentioned in Note 2.2.1 (Section 2.2).

[19] In particular by (A.2), once understood the same symbol for corresponding dimensional and non-dimensional entities, as declared in Note 3.4.1 above.

For $M_\star \to 0$ the non-dimensional form of the semi-discrete system (3.72), which is derived in Section A.2 for ease of presentation, reads (as usual, $i \in \mathcal{I}$):

$$\begin{cases} 2\mu_i \dfrac{d}{dt}(\rho_i) = M_\star^{-1} \Psi_{sd}^{(-1)} + \hat{\Psi}_{sd}^{(0)} \\[4mm] 2\mu_i \dfrac{d}{dt}(\rho_i u_i) = M_\star^{-2} \Theta_{sd}^{(-2)} + M_\star^{-1} \Theta_{sd}^{(-1)} + \hat{\Theta}_{sd}^{(0)} \end{cases} \tag{3.80}$$

where the relevant coefficients are defined in the aforementioned section.[20]

Asymptotic analysis

Following [42], solutions are sought to the continuous problem (3.78) and the semi-discrete one (3.80) in the nearly-incompressible limit (*i.e.* for $M_\star \to 0$), in the form of asymptotic expansions in power of $M_\star$, namely:

$$\begin{cases} \rho(x, t) = \rho_0(x, t) + M_\star \rho_1(x, t) + M_\star^2 \rho_2(x, t) + \cdots \\ u(x, t) = u_0(x, t) + M_\star u_1(x, t) + M_\star^2 u_2(x, t) + \cdots \\ p(x, t) = p_0(x, t) + M_\star p_1(x, t) + M_\star^2 p_2(x, t) + \cdots \end{cases} \tag{3.81}$$

for the continuous system and:

$$\begin{cases} \rho_i(t) = \rho_{0i}(t) + M_\star \rho_{1i}(t) + M_\star^2 \rho_{2i}(t) + \cdots \\ u_i(t) = u_{0i}(t) + M_\star u_{1i}(t) + M_\star^2 u_{2i}(t) + \cdots \\ p_i(t) = p_{0i}(t) + M_\star p_{1i}(t) + M_\star^2 p_{2i}(t) + \cdots \end{cases} \tag{3.82}$$

for the semi-discrete one. All the entities appearing in the expansions above are supposed to be regular enough (for any further manipulations).

Note 3.4.2. By virtue of the considered barotropic state law (1.3), it is possible to derive some relations between the coefficients $\rho_k(\cdot, \cdot)$ and $p_h(\cdot, \cdot)$ appearing in (3.81). For instance,

$$\begin{aligned} p = p(\rho) &= p(\rho_0 + M_\star \rho_1 + \cdots) \\ &= p(\rho_0) + M_\star a^2(\rho_0) \rho_1 + \cdots \end{aligned}$$

and therefore:

$$\begin{cases} p_0 = p(\rho_0) \\ p_1 = a^2(\rho_0) \rho_1 . \end{cases}$$

---

[20] In particular by (A.12) and (A.15), once understood the same symbol for corresponding dimensional and non-dimensional entities, as declared in Note 3.4.1 above.

Similar considerations can be introduced for the coefficients $\rho_{ki}(\cdot)$ and $p_{hi}(\cdot)$ in (3.82); thus, for instance, the following relations hold as well:

$$\begin{cases} p_{0i} = p(\rho_{0i}) \\ p_{1i} = a^2(\rho_{0i})\, \rho_{1i} \,. \end{cases} \tag{3.83}$$

Note 3.4.3. It is possible to exploit the equalities in (3.82) in order to also expand the Roe averages. For instance, the coefficient a_{0ij} appearing in the following expansion of a_{ij}:

$$a_{ij}(t) = a_{0ij}(t) + M_\star\, a_{1ij}(t) + M_\star^2\, a_{2ij}(t) + \cdots$$

can be obtained as follows:

- if $\Delta^{ij}\rho = 0$ then $\rho_i = \rho_j = \bar{\rho}$ and, according to (3.62) and (3.82), the following relation holds:

$$\begin{aligned} a_{ij} &= a\,(\bar{\rho}) \\ &= a\,(\bar{\rho}_0 + M_\star\, \bar{\rho}_1 + \cdots) \\ &= a\,(\bar{\rho}_0) + M_\star\, \frac{da}{d\rho}\,(\bar{\rho}_0)\, \bar{\rho}_1 + \cdots \end{aligned}$$

Then, clearly:

$$a_{0ij} = a\,(\bar{\rho}_0)$$

- if $\Delta^{ij}\rho \neq 0$ then, by only considering the zero-order terms in the expansion of the following equality (which is directly obtained from (3.62)): $\Delta^{ij} p = a_{ij}^2\, \Delta^{ij}\rho$, the subsequent relation is obtained: $\Delta^{ij} p_0 = a_{0ij}^2\, \Delta^{ij}\rho_0$. The coefficient a_{0ij} is positive since a_{ij} is positive (by definition) and $a_{ij}\,(M_\star \to 0) \to a_{0ij}$. Hence, $\Delta^{ij} p_0 = 0 \Leftrightarrow \Delta^{ij}\rho_0 = 0$. As a consequence,

  - if $\Delta^{ij}\rho_0 \neq 0$, then:

$$\begin{aligned} a_{ij} &= \left(\frac{\Delta^{ij} p}{\Delta^{ij}\rho} \right)^{\frac{1}{2}} \\ &= \left(\frac{\Delta^{ij} p_0}{\Delta^{ij}\rho_0} \right)^{\frac{1}{2}} \left(1 + M_\star \frac{\Delta^{ij} p_1}{\Delta^{ij} p_0} + \cdots \right)^{\frac{1}{2}} \left(1 + M_\star \frac{\Delta^{ij}\rho_1}{\Delta^{ij}\rho_0} + \cdots \right)^{-\frac{1}{2}} \\ &= \left(\frac{\Delta^{ij} p_0}{\Delta^{ij}\rho_0} \right)^{\frac{1}{2}} + M_\star\, \frac{1}{2} \left(\frac{\Delta^{ij} p_0}{\Delta^{ij}\rho_0} \right)^{\frac{1}{2}} \left(\frac{\Delta^{ij} p_1}{\Delta^{ij} p_0} - \frac{\Delta^{ij}\rho_1}{\Delta^{ij}\rho_0} \right) + \cdots \end{aligned}$$

    In this case:

$$a_{0ij} = \left(\frac{\Delta^{ij} p_0}{\Delta^{ij}\rho_0} \right)^{\frac{1}{2}}$$

– if $\Delta^{ij} \rho_0 = 0$, then by exploiting the same kind of linearization as above, the following expression is obtained:

$$a_{0ij} = \left(\frac{\Delta^{ij} p_k}{\Delta^{ij} \rho_k} \right)^{\frac{1}{2}}$$

where k denotes the first integer such that $\Delta^{ij} \rho_k \neq 0$.

Of course, once defined the relevant coefficients appearing in the expansion of the Roe averages, it is possible to expand all the derived entities as well. For instance, once noticed that (as reminded above, $a_{0ij} > 0$):

$$a_{ij}^{-1} = a_{0ij}^{-1} \left(1 + M_\star \frac{a_{1ij}}{a_{0ij}} + \cdots \right)^{-1} = a_{0ij}^{-1} \left(1 - M_\star \frac{a_{1ij}}{a_{0ij}} + \cdots \right)$$

the parameter M_{ij} introduced in (A.13) (Section A.2) and reported below for the sake of clarity:

$$M_{ij} := \frac{u_{ij}}{a_{ij}}$$

admits the following asymptotic expression:

$$M_{ij} = \frac{u_{0ij}}{a_{0ij}} + M_\star \left(\frac{u_{1ij}}{a_{0ij}} - \frac{u_{0ij} a_{1ij}}{a_{0ij}^2} \right) + \cdots$$

Hence, in particular, $M_{0ij} = \dfrac{u_{0ij}}{a_{0ij}}$.

By exploiting the aforementioned expansions, it is possible to state the following:

Proposition 3.4.4. *For $M_\star \to 0$, the pressure associated with the solution of the continuous problem (3.78) is of the form:*

$$p(x, t) = \bar{p}_0(t) + M_\star \, \bar{p}_1(t) + M_\star^2 \, p_2(x, t) + \cdots \qquad (3.84)$$

while that one relative to the semi-discrete problem (3.80) admits the following representation:

$$p_i(t) = \tilde{p}_0(t) + M_\star \, p_{1i}(t) + \cdots \qquad (3.85)$$

Proof. The proof is reported in Section A.3, for ease of presentation. $\square$

By comparing the expansions (3.84) and (3.85) it is clear that, in the nearly-incompressible limit, the semi-discrete solution admits pressure variations in space higher than those associated with the continuous one. As a consequence, for $M_\star \to 0$ the discrete schemes based on the proposed Roe numerical flux (3.73)-(3.75) (*i.e.* (3.64)-(3.68)) may provide a numerical solution remarkably different from the continuous one. In other words, the accuracy of the considered compressible solvers can be dramatically reduced when the flow tends to become (even locally) nearly-incompressible.

Note 3.4.5. An asymptotic behaviour of the same kind of that one described by the expansions (3.84) and (3.85) is obtained in [42], when considering the compressible Euler equations coupled with a perfect gas state law.

3.4.2. A brief introduction to preconditioning techniques for the low speed Euler and Navier-Stokes equations

The considered preconditioning techniques originate from the "artificial compressibility" method proposed by Chorin [18] for determining a steady-state solution to the incompressible Navier-Stokes equations. When considering the two-dimensional incompressible equations written in terms of the so-called "primitive" variables p, u_1 and u_2 (where, of course, u_1 and u_2 denote the components of the velocity vector), the continuity equation reads:

$$\partial_{x_1} u_1 + \partial_{x_2} u_2 = 0 .$$

Then, by following the artificial compressibility method a fictitious pressure time-derivative is added to the above equation, as follows:

$$\kappa^{-1} \partial_t p + \partial_{x_1} u_1 + \partial_{x_2} u_2 = 0$$

where κ is a constant. Once completed the set of the governing equations by also considering the proper momentum balance, the original and the modified system only differ from each other as for the time-derivative term, which in the former case reads:

$$\partial_t \begin{pmatrix} 0 \\ u_1 \\ u_2 \end{pmatrix}$$

while in the latter one can be expressed as follows:

$$\mathbf{P}_{\text{Chorin}}^{-1} \cdot \partial_t \begin{pmatrix} p \\ u_1 \\ u_2 \end{pmatrix}$$

with:

$$\mathbf{P}_{\text{Chorin}}^{-1} := \begin{pmatrix} \kappa^{-1} & 0 & 0 \\ 0 & 1 & 0 \\ 0 & 0 & 1 \end{pmatrix}. \tag{3.86}$$

By introducing the aforementioned pressure time-derivative term, the decoupling between the pressure and the velocity field, which represents an important issue for the numerical discretization of the incompressible equations, is avoided and the hyperbolicity of the governing system is restored. However, the modified system is not consistent in time and therefore the considered formulation can only be exploited in order to march towards a steady-state solution, hopefully by guaranteeing a stable and efficient convergence through the definition of the parameter κ in $\mathbf{P}_{\text{Chorin}}^{-1}$.

Note 3.4.6. In consideration of the fact that the artificial compressibility formulation results in a hyperbolic system of equations, its discretization is carried out in [5] by exploiting classical techniques conceived for compressible flows (namely a finite volume method based on upwind schemes and Riemann solvers). In particular, a linearized implicit time-advancing strategy is defined in which the parameter κ, originally associated with the fictitious pressure time-derivative, appears in such a way that the consistency in time is preserved. Hence, the considered scheme can be exploited for unsteady simulations as well.

The basic idea of pre-multiplying the time-derivative term by a suitable matrix gave birth to a class of numerical methods designed for improving the convergence of the compressible Euler and Navier-Stokes equations to a steady-state. In particular, the time-derivatives are modulated (again, at the cost of loosing the consistency in time) in order to achieve a stable and efficient time-marching; in this spirit, the matrix under consideration is referred to as a preconditioner. The resulting schemes are generally referred to as "pseudo-unsteady" methods (see *e.g.* [75]) and the considered preconditioning technique is often indicated as "iterative preconditioning" (see *e.g.* [112]).

The numerical technique under consideration has been also exploited for the numerical simulation of flows in which there is a significant discrepancy between the convective and the acoustic speeds (*i.e.* time-scales). In particular, it has been used for reducing the numerical stiffness of compressible solvers dealing with flows at low Mach numbers. In this context, a well-known preconditioner has been put forward by Turkel [100]. Once introduced a state vector $\mathbf{z}$ such that $d\mathbf{z} = ((\rho a)^{-1} dp, du_1, du_2, a c_p^{-1} ds)^T$ where s indicates the entropy per unit mass of the fluid and c_P its specific heat at constant pressure (a denoting the sound speed), the expression of the considered preconditioner which appears in the compressible equations written in terms of $\mathbf{z}$ reads [101]:

$$\mathbf{P}_{\mathrm{Turkel}}^{-1} := \begin{pmatrix} \dfrac{1}{\beta^2} & 0 & 0 & \delta \\[2ex] \dfrac{\alpha u_1}{a\,\beta^2} & 1 & 0 & 0 \\[2ex] \dfrac{\alpha u_2}{a\,\beta^2} & 0 & 1 & 0 \\[2ex] 0 & 0 & 0 & 1 \end{pmatrix} \tag{3.87}$$

where α, β and δ indicate suitable non-dimensional parameters. In particular β is chosen of the order of $\tilde{M}$, where $\tilde{M}$ denotes the characteristic Mach number of the flow field, for the considered preconditioning to be effective. For $\alpha = 0$ and $\delta = 1$ the matrix (3.87) reduces to another classical preconditioner introduced by Choi and Merkle [17] while for $\delta = 0$ it clearly generalizes that matrix of Chorin (3.86), by also altering the time-derivative appearing in the momentum balance.

The iterative preconditioning has been also introduced when considering upwind schemes like, for instance, the Roe scheme. The upwinding strategy, in particular, can be applied to the preconditioned formulation (see e.g. [109]). The resulting scheme, in general, may be not consistent in time. However, by confining the effect of the preconditioner within a portion of the numerical scheme which does not affect its consistency in time (as, for instance, that one associated with the upwind component of the Roe numerical flux when dealing with the corresponding scheme), it is possible to exploit the resulting scheme for unsteady simulations as well. A time-consistent preconditioning strategy is defined, in particular, in [42] and [112], which is recalled in the following Section 3.4.3.

It may be worth mentioning that, as an alternative to the aforementioned approach, a dual time-step strategy is usually adopted in order to overcome the time-consistency problem (see e.g. [23, 58, 73] and [102]). More in detail, by starting from the following system:

$$\partial_t \mathbf{z} + \mathbf{r}\,(\mathbf{z}) = \mathbf{0}$$

in which $\mathbf{r}$ denotes the steady-state residual as a function of the chosen state vector $\mathbf{z}$, an additional term is added, as follows:

$$\mathbf{P}^{-1} \cdot \partial_\tau \mathbf{z} + \partial_t \mathbf{z} + \mathbf{r}\,(\mathbf{z}) = \mathbf{0}$$

where τ denotes a fictitious time and $\mathbf{P}^{-1}$ is a suitable matrix. More in detail, by advancing the solution of the latter system with respect to τ up to a steady-state,[21] the solution of the former one is recovered; the consistency in time is clearly preserved. The matrix $\mathbf{P}^{-1}$ is designed for optimizing the aforementioned convergence and therefore represents a preconditioner (in the sense of the iterative preconditioning).

3.4.3. Preconditioning of the Roe numerical flux

As pointed out in Note 3.4.5 above, the asymptotic behaviours obtained for the perfect gas state law and for a generic barotropic state law are similar to each other. Hence, the preconditioning technique proposed in [42] is also considered for the barotropic case. Basically, it consists in replacing the Roe flux (3.73)-(3.75) with the following expression:

$$\boldsymbol{\phi}^{(x)\mathrm{ROE},p}\left(\mathbf{q}_i^{(x)}, \mathbf{q}_j^{(x)}, \hat{\boldsymbol{v}}_{ij}\right) := \boldsymbol{\phi}_{ij}^{(x)\mathrm{ROE},p}, \quad j \in \pi_i \qquad (3.88)$$

where:

$$\boldsymbol{\phi}_{ij}^{(x)\mathrm{ROE},p} := \boldsymbol{\phi}_{c,ij}^{(x)\mathrm{ROE}} + \boldsymbol{\phi}_{u,ij}^{(x)\mathrm{ROE},p} \qquad (3.89)$$

$$\boldsymbol{\phi}_{u,ij}^{(x)\mathrm{ROE},p} := \mathbf{D}_{ij}^{(x),p} \cdot \Delta^{ij}\mathbf{q}^{(x)} $$

$$\mathbf{D}_{ij}^{(x),p} := -\frac{1}{2}\left(\mathbf{P}_{ij}^{(x)}\right)^{-1} \cdot \left|\mathbf{P}_{ij}^{(x)} \cdot \left(s_{ij}\,\tilde{\mathbf{J}}_{ij}^{(x)}\right)\right| \qquad (3.90)$$

and $\boldsymbol{\phi}_{c,ij}^{(x)\mathrm{ROE}}$ in (3.89) is defined by (3.74). It should be noticed that only the upwind component of the flux function is modified, by means of the preconditioning matrix $\mathbf{P}_{ij}^{(x)}$ defined below. Let $\mathbf{w}_p^{(x)}$ denote the following basic-1D primitive state vector:

$$\mathbf{w}_p^{(x)} := \begin{pmatrix} p \\ u \end{pmatrix}.$$

Then, the following matrix:

$$\mathbf{P}_{\mathbf{q}}^{(x)} := \partial_{\mathbf{w}_p^{(x)}}\mathbf{q}^{(x)} \cdot \mathbf{P}_{\mathbf{w}_p}^{(x)} \cdot \partial_{\mathbf{q}^{(x)}}\mathbf{w}_p^{(x)} \qquad (3.91)$$

where $\mathbf{q}^{(x)}$ denotes the basic-1D conservative state vector (2.15) and $\mathbf{P}_{\mathbf{w}_p}^{(x)}$ is defined as follows:

$$\mathbf{P}_{\mathbf{w}_p}^{(x)} := \begin{pmatrix} \beta^2 & 0 \\ 0 & 1 \end{pmatrix}, \quad \beta = \mathrm{const} \qquad (3.92)$$

---

[21] Of course, in a practical set-up only a limited number of time-advancing steps are performed.

defines a function of u, namely:

$$\mathbf{P}_{\mathbf{q}}^{(x)}(u) = \mathbf{I} + \left(\beta^2 - 1\right) \begin{pmatrix} 1 & 0 \\ u & 0 \end{pmatrix}. \tag{3.93}$$

The matrix $\mathbf{P}_{ij}^{(x)}$ is finally defined by evaluating (3.93) in correspondence of the proper Roe average, as follows:

$$\mathbf{P}_{ij}^{(x)} := \mathbf{P}_{\mathbf{q}}^{(x)}(u = u_{ij}) \tag{3.94}$$

It is worth remarking that the matrix $\mathbf{P}_{ij}^{(x)} \cdot \left(s_{ij}\,\tilde{\mathbf{J}}_{ij}^{(x)}\right)$ appearing in the pre-conditioned upwind component (3.90) is diagonalizable with real eigenvalues (see Section A.4 for details) and therefore the operator $|\cdot|$, defined by (2.1), can be rightfully applied.

Note 3.4.7. The numerical flux (3.88)-(3.90) is generally referred to as the Roe-Turkel numerical flux (see *e.g.* [42]). Indeed, the matrix (3.92) can be derived from the 1D counterpart of the preconditioner of Turkel (3.87) for $\alpha = \delta = 0$, by a change of variables.[22]

Note 3.4.8. The preconditioner is usually introduced by considering a quasi-linear formulation, consistently with the fact that regular solutions (*e.g.* without shocks) are investigated in the nearly-incompressible limit. Several sets of independent variables can be chosen in order to derive the preconditioner (see *e.g.* [102] and [110]).[23] For the present case, the preconditioner is introduced in terms of the primitive variables through the matrix (3.92) and then it is converted to the conservative variables by means of the expression (3.91). The specific form of the adopted state law comes into play at this point of the derivation; the expression (3.93), in particular, is valid for a (generic) barotropic state law.

In order to assess the effects the considered preconditioning technique produces on the asymptotic semi-discrete solution, the following semi-discrete system is considered:

$$\mu_i \frac{d}{dt} \mathbf{q}_i^{(x)} + \sum_{j \in \pi_i} \boldsymbol{\phi}^{(x)\text{ROE},p}\left(\mathbf{q}_i^{(x)}, \mathbf{q}_j^{(x)}, \hat{\boldsymbol{v}}_{ij}\right) = \mathbf{0} \;,\; i \in \mathcal{I} \tag{3.95}$$

where the numerical flux $\boldsymbol{\phi}^{(x)\text{ROE},p}\left(\cdot, \cdot, \cdot\right)$ is given by (3.88)-(3.90). For $M_\star \to 0$ the non-dimensional form of the semi-discrete system (3.95),

---

[22] For $\alpha = \delta = 0$, the preconditioner of Turkel reduces to that one of Chorin, *i.e.* (3.86).

[23] The specific choice affects the convergence to a steady-state and the accuracy of the numerical solutions for low Mach number steady and unsteady flows [102].

which is derived in Section A.4 for ease of presentation, reads (as usual, $i \in \mathcal{I}$):

$$
\begin{cases}
2\mu_i \dfrac{d}{dt}(\rho_i) = M_\star^{-1} \Psi_{sd,p}^{(-1)} + \hat{\Psi}_{sd,p}^{(0)} \\[2ex]
2\mu_i \dfrac{d}{dt}(\rho_i u_i) = M_\star^{-2} \Theta_{sd,p}^{(-2)} + M_\star^{-1} \Theta_{sd,p}^{(-1)} + \hat{\Theta}_{sd,p}^{(0)}
\end{cases}
\tag{3.96}
$$

where the relevant coefficients are defined in the aforementioned section.[24]

The expansion (3.96) is obtained by assuming that the parameter β in (3.92) is formally of the order of the unity. However, by following [42], the parameter β is hereafter assumed of the order of the characteristic Mach number $M_\star$, namely:

$$
\beta = \beta_{\text{ref}} M_\star
\tag{3.97}
$$

where β_{ref} is a given constant of the order of the unity. The position (3.97) renders the considered preconditioning technique effective, as shown in the sequel. First of all, the fact that now β explicitly introduces the factor $M_\star$ leads to a non-dimensional system different from (3.96). In particular, it is possible to show that for $M_\star \to 0$ the non-dimensional form of (3.95) now reads:

$$
\begin{cases}
2\mu_i \dfrac{d}{dt}(\rho_i) = M_\star^{-2} \check{\Psi}_{sd,p}^{(-2)} + M_\star^{-1} \check{\Psi}_{sd,p}^{(-1)} + \ddot{\Psi}_{sd,p}^{(0)} \\[2ex]
2\mu_i \dfrac{d}{dt}(\rho_i u_i) = M_\star^{-2} \check{\Theta}_{sd,p}^{(-2)} + M_\star^{-1} \check{\Theta}_{sd,p}^{(-1)} + \ddot{\Theta}_{sd,p}^{(0)} .
\end{cases}
\tag{3.98}
$$

The definition of the coefficients appearing in (3.98) is not reported here because inessential to the present purposes. However, it should be noticed that the coefficient $\check{\Psi}_{sd,p}^{(-2)}$ in (3.98) has no counterpart in (3.96). Indeed, it derives from the position (3.97), by a mechanism of the type of that one mentioned in Section A.5. The behaviour of the the system (3.98) in the nearly-incompressible limit is described by the following:

Proposition 3.4.9. *For $M_\star \to 0$, the pressure associated with the solution of the semi-discrete problem (3.98) admits the following representation:*

$$
p_i(t) = \hat{p}_0(t) + M_\star \, \hat{p}_1(t) + M_\star^2 \, p_{2i}(t) + \cdots
\tag{3.99}
$$

---

[24] In particular by (A.31) and (A.34), once understood the same symbol for corresponding dimensional and non-dimensional entities, as declared in Note 3.4.1 (Section 3.4.1).

Proof. The proof is reported in Section A.6, for ease of presentation.   □

By comparing the expansions (3.99) and (3.84) it is clear that, in the nearly-incompressible limit, the solution associated with the preconditioned semi-discrete formulation exhibits a behaviour which is qualitatively similar to that of the continuous one. This should result, in principle, in a more accurate discrete solution for $M_\star \to 0$, as confirmed by the numerical results reported in the following Section 3.4.4.

Note 3.4.10. It is straightforward to extend the considered preconditioning technique to the augmented-1D systems defined in Section 2.2.4. To the purpose, the following numerical flux function is introduced:

$$\boldsymbol{\phi}^{(A)\text{ROE},p}\left(\mathbf{q}_i^{(A)}, \mathbf{q}_j^{(A)}, \hat{\mathbf{v}}_{ij}\right) := \boldsymbol{\phi}_{ij}^{(A)\text{ROE},p}, \quad j \in \pi_i \tag{3.100}$$

with:

$$\boldsymbol{\phi}_{ij}^{(A)\text{ROE},p} := \boldsymbol{\phi}_{c,ij}^{(A)\text{ROE}} + \boldsymbol{\phi}_{u,ij}^{(A)\text{ROE},p}$$

$$\boldsymbol{\phi}_{u,ij}^{(A)\text{ROE},p} := \mathbf{D}_{ij}^{(A),p} \cdot \Delta^{ij}\mathbf{q}^{(A)}$$

$$\mathbf{D}_{ij}^{(A),p} := -\frac{1}{2}\left(\mathbf{P}_{ij}^{(A)}\right)^{-1} \cdot \left|\mathbf{P}_{ij}^{(A)} \cdot \left(s_{ij}\,\tilde{\mathbf{J}}_{ij}^{(A)}\right)\right| \tag{3.101}$$

where $\boldsymbol{\phi}_{c,ij}^{(A)\text{ROE}}$ is formally given by (3.66) and:

- if $\mathbf{q}^{(A)}$ is defined by (2.19) then $\mathbf{f}^{(A)}$ is given by (2.20), $\tilde{\mathbf{J}}_{ij}^{(A)}$ is given by (3.60) and the preconditioner reads:

$$\mathbf{P}_{ij}^{(A)} := \mathbf{I} + \left(\beta^2 - 1\right) \begin{pmatrix} 1 & 0 & 0 \\ u_{ij} & 0 & 0 \\ \xi_{ij} & 0 & 0 \end{pmatrix}$$

- if $\mathbf{q}^{(A)}$ is given by (2.23) then $\mathbf{f}^{(A)}$ is given by (2.24), $\tilde{\mathbf{J}}_{ij}^{(A)}$ is defined analogously to (3.59) and the preconditioner reads:

$$\mathbf{P}_{ij}^{(A)} := \mathbf{I} + \left(\beta^2 - 1\right) \begin{pmatrix} 1 & 0 & 0 & 0 \\ u_{ij} & 0 & 0 & 0 \\ \xi_{ij} & 0 & 0 & 0 \\ \eta_{ij} & 0 & 0 & 0 \end{pmatrix}. \tag{3.102}$$

3.4.4. Numerical results

Benchmark

A quasi-1D, inviscid, barotropic flow within a duct having variable cross-sectional area $A = A(x)$ (*e.g.* a convergent-divergent nozzle) is considered. Let $\mathbf{q}^{(x)}$ and $\mathbf{f}^{(x)}$ be defined by (2.15) and (2.16), respectively. The relevant mass and momentum balances read (compare with (2.18)):

$$\partial_t \, \mathbf{q}^{(x)} + \partial_x \, \mathbf{f}^{(x)} = \mathbf{s}^{(x)} \left(\mathbf{q}^{(x)} \right) \tag{3.103}$$

with:

$$\mathbf{s}^{(x)} \left(\mathbf{q}^{(x)} \right) := -\frac{\mathrm{d}}{\mathrm{d}x} \ln \left(A(x) \right) \begin{pmatrix} \rho \, u \\ \rho \, u^2 \end{pmatrix}.$$

As declared at the beginning of Section 3.4, a smooth solution to (3.103) is sought in the present context. In particular, the steady, nearly-incompressible solution to (3.103) described below is considered in order to define a benchmark for the proposed preconditioning strategy.

Let $M_\star \ll 1$ denote the characteristic Mach number of a flow field exhibiting a roughly constant density:

$$\rho \approx \rho_\infty \tag{3.104}$$

where the subscript ∞ hereafter refers to the inlet conditions, associated with the section located at $x = x_{min} \in [x_{min}, x_{max}]$. In consideration of (3.104), the conservation of the mass approximately reduces to the following relation:

$$u \, A \approx u_\infty \, A_\infty \tag{3.105}$$

Moreover, by invoking the well-known Bernoulli theorem (for incompressible, non-dissipative steady flows) [88], the momentum balance can be approximated as follows:

$$p + \frac{1}{2} \, \rho_\infty \, u^2 \approx p_\infty + \frac{1}{2} \, \rho_\infty \, u_\infty^2 . \tag{3.106}$$

Then, once defined:

$$\alpha(x) := \frac{A(x)}{A_\infty} \tag{3.107}$$

it is possible to respectively recast (3.105) and (3.106) as follows:

$$\frac{u}{u_\infty} \approx \alpha^{-1} \tag{3.108}$$

$$\frac{p}{p_\infty} \approx 1 + \frac{1}{2} \, \frac{\rho_\infty \, u_\infty^2}{p_\infty} \left(1 - \alpha^{-2} \right) . \tag{3.109}$$

The relations (3.108) and (3.109) provide a nearly-exact, steady solution to (3.103) (they tend to be exact for $M_\star \to 0$) which can be exploited for validating the discrete scheme (3.113), based on the preconditioned numerical flux (3.88)-(3.90).

As far as the variation of the cross-sectional area is concerned, the following definition is adopted, in particular, for $\alpha(x)$:

$$\alpha(x) := \begin{cases} 1 & \text{if } x_{\min} \le x \le x_1 \\ 1 - \sigma \left(1 - \cos \left(2\pi \, \dfrac{x - x_1}{x_2 - x_1} \right) \right) & \text{if } x_1 < x < x_2 \\ 1 & \text{if } x_2 \le x \le x_{\max} \end{cases} \quad (3.110)$$

where x_1, x_2 and $0 < \sigma < 1/2$ are adjustable parameters. The function (3.110) represents a sinusoidal reduction of the cross-sectional area between x_1 and x_2. In particular, the minimum value of α is given by:

$$\alpha_{\min} := 1 - 2\sigma \quad (3.111)$$

and it is obtained for $x = (x_1 + x_2)/2$.

The considered benchmark is summarized in Table 3.6. In this table, κ, $\varkappa$ and γ characterize an instance of the convex barotropic state law (2.66) which introduces, in particular, a constant sound speed $\tilde{a} = \sqrt{\kappa} = 10^3$ for the flow. The variation of the cross-sectional area $A/A_\infty = \alpha$ is shown in Figure 3.21. In consideration of (3.108), the maximum value of u is given by $u \approx \alpha_{\min}^{-1} \approx 1.05$ and therefore $M_\star = 10^{-3}$ can be regarded to as a characteristic Mach number of the considered flow.

Table 3.6. Considered benchmark.

Benchmark	κ	$\varkappa$	γ	ρ_∞	u_∞	$x_{\min}$	x_1	x_2	$x_{\max}$	σ
BN	10^6	1	0	1	1	-2000	-1000	1000	2000	$2.5 \cdot 10^{-2}$

Discrete scheme, initial and boundary conditions

By starting from the system (3.72), the following semi-discrete formulation is introduced, based on the proposed preconditioned numerical flux (3.88)-(3.90):

$$\mu_i \frac{d}{dt} \mathbf{q}_i^{(x)} + \sum_{j \in \pi_i} \phi^{(x)\text{ROE},p} \left(\mathbf{q}_i^{(x)}, \mathbf{q}_j^{(x)}, \hat{\mathbf{v}}_{ij} \right) = \mathbf{s}_i^{(x)}, \quad i \in \mathcal{I} \quad (3.112)$$

where:

$$\mathbf{s}_i^{(x)} \approx \int_{C_i} \mathbf{s}^{(x)} \, dx .$$

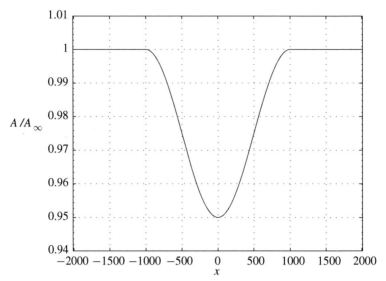

Figure 3.21. Variation of the cross-sectional area for the test-case under consideration.

In particular, the following definition is adopted:

$$\mathbf{s}_i^{(x)} := \gamma_i \begin{pmatrix} \rho_i \, u_i \\ \rho_i \, u_i^2 \end{pmatrix}$$

with:

$$\gamma_i := \ln \left(\frac{A(x_{i-1/2})}{A(x_{i+1/2})} \right) = \ln \left(\frac{\alpha(x_{i-1/2})}{\alpha(x_{i+1/2})} \right).$$

Then, in the spirit of the basic explicit scheme (3.10), the following discretization is considered:

$$\mathbf{q}_i^{(x)n+1} = \mathbf{q}_i^{(x)n} + \frac{\delta^n t}{\mu_i} \left(-\sum_{j \in \pi_i} \boldsymbol{\phi}^{(x)\mathrm{ROE},p} \left(\mathbf{q}_i^{(x)n}, \mathbf{q}_j^{(x)n}, \hat{\mathbf{v}}_{ij} \right) + \mathbf{s}_i^{(x)n} \right), \quad i \in \mathcal{I} \tag{3.113}$$

The following uniform field is chosen as IC:

$$\mathbf{q}_i^{(x)0} = \mathbf{q}_\infty^{(x)}, \quad i \in \mathcal{I} \tag{3.114}$$

where:

$$\mathbf{q}_\infty^{(x)} := \begin{pmatrix} \rho_\infty \\ \rho_\infty \, u_\infty \end{pmatrix} \tag{3.115}$$

and both ρ_∞ and u_∞ are introduced in the previous paragraph.

As far as the BCs are concerned, a Dirichlet-like inlet BC is enforced by defining the fictitious state vector $\mathbf{q}_0^{(x)n}$ as follows:

$$\mathbf{q}_0^{(x)n} = \mathbf{q}_\infty^{(x)} , \quad n = 0, 1, 2, \ldots \tag{3.116}$$

while the following transmissive BC is adopted at the outlet:

$$\mathbf{q}_{N_c+1}^{(x)n} = \mathbf{q}_{N_c}^{(x)n} , \quad n = 0, 1, 2, \ldots \tag{3.117}$$

A remark on linear stability

The explicit discrete scheme (3.113) is subjected to a CFL-like constraint of the type of (3.29), as the basic explicit scheme (3.10) from which it is derived. Independently of $s_i^{(x)}$, which accounts for the specific geometry of the duct, it makes sense to investigate the effect the preconditioning strategy has on the wave structure of the linearized problem and in particular on the corresponding maximum wave speed s_{max}, since it directly affects the CFL-like constraint. To the purpose, some considerations are reported below; a rather informal presentation is adopted for the sake of simplicity.

An estimate of s_{max} can be obtained by considering the spectral radius of the matrix $\mathbf{D}_{ij}^{(x),p}$ defined in (3.90). In particular, it is possible to linearize the flow field in the neighbourhood of a certain point $\tilde{\mathbf{q}}^{(x)}$ and, by virtue of the property (RM2) reported in Section 3.3.1, to evaluate the spectral radius of $\mathbf{D}_{ij}^{(x),p}$ in correspondence of $\tilde{\mathbf{q}}^{(x)}$ (see *e.g.* [8]). Straightforward algebraic manipulations (not reported here for the sake of conciseness) show that:

$$M_\star \ll 1 \Rightarrow \mathbf{D}_{ij}^{(x),p} \approx \begin{pmatrix} O\left(M_\star^{-1}\tilde{a}\right) & O\left(M_\star^{-2}\right) \\ O\left(\tilde{a}^2\right) & O\left(M_\star^{-1}\tilde{a}\right) \end{pmatrix}$$

where $\tilde{a}$ denote the characteristic sound speed associated with $\tilde{\mathbf{q}}^{(x)}$. Clearly, $s_{max} = O\left(M_\star^{-1}\tilde{a}\right)$ for the present, preconditioned case. On the other hand, the non-preconditioned case can be analysed exactly in the same manner (in particular, by simply choosing $\beta^2 = 1$ where appropriate), thus obtaining:

$$M_\star \ll 1 \Rightarrow \mathbf{D}_{ij}^{(x)} \approx \begin{pmatrix} O\left(\tilde{a}\right) & O\left(M_\star\right) \\ O\left(M_\star\tilde{a}^2\right) & O\left(\tilde{a}\right) \end{pmatrix}.$$

In this case, $s_{max} = O\left(\tilde{a}\right)$ (as already noticed, for instance, in Section 3.3.2). On the basis of the aforementioned analysis, it is clear that the largest wave speed increases of $O\left(M_\star^{-1}\right) \gg 1$ when switching the preconditioning technique on. In other words, in consideration of the CFL-like constraint (3.29), it should be necessary to reduce the time-step of $O\left(M_\star\right) \ll 1$ in order to keep the explicit time-advancing stable:

$$\tau_{\text{prec}} = O\left(M_\star\right) \cdot \tau_{\text{noprec}} . \qquad (3.118)$$

As a result, the considered (explicit preconditioned) discrete scheme (3.113) should exhibit very severe efficiency limitations. This point is confirmed by the numerical results reported in the following paragraph.

Test cases

The considered test-cases are summarized in Table 3.7.

Table 3.7. Considered test-cases for the discrete scheme (3.113), based on the preconditioned numerical flux (3.88)-(3.90).

Test-case	Benchmark	μ	β^2	τ
ER-NOPREC	BN	10	1	$5 \cdot 10^{-3}$
ER-PREC	BN	10	10^{-6}	$5 \cdot 10^{-6}$

In particular:

- the x-domain $[x_{\min}, x_{\max}]$ is uniformly discretized (with cells having size μ) for both cases;
- for the test-case ER-NOPREC the proposed preconditioning strategy is not activated (indeed for $\beta^2 = 1$ the preconditioner (3.93) reduces to the identity matrix and therefore it does not modify the numerical flux function). Conversely, the test-case ER-PREC exploits the preconditioning strategy at hand, by choosing β_{ref} in (3.97) exactly equal to 1;
- the time-step τ for the test-case ER-NOPREC is chosen equal to that one associated with the test-case ER1-2 in Table 3.4 (Section 3.3.2), since both the considered (non-preconditioned) test-cases are based on the same state law as well as the same space discretization. The value which is chosen for the test-case ER-PREC, instead, represents the maximum time-step which can be adopted, as a matter of fact, in order to obtain a stable time-advancing (up to the benchmark steady-state).

The behaviour of the corresponding numerical solutions is shown in Figures 3.22 and 3.23. It should be noticed that:

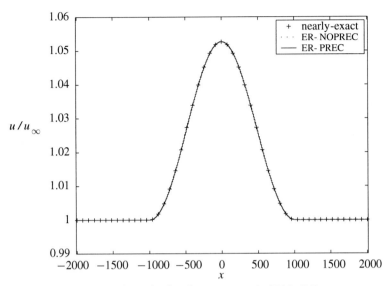

Figure 3.22. Approximation of u for the test-cases in Table 3.7.

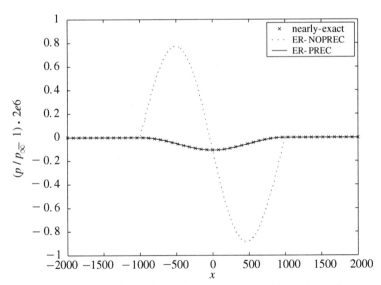

Figure 3.23. Approximation of p for the test-cases in Table 3.7. The pressure variation on the y-axis is scaled for ease of readability.

- the approximation of u does not suffer from the accuracy problems related to the low Mach number flow (its main driver being the area variation, according to (3.105)), while the approximation of p exhibits the problems highlighted in Proposition 3.4.4 (Section 3.4.1). The proposed preconditioning strategy, however, seems to effectively counteract these problems, as shown in Figure 3.23;

- the time-steps reported in Table 3.7 clearly satisfies the relation (3.118), thus supporting the considerations regarding the stability of the considered scheme which are introduced in the previous paragraph. The extremely small time-step required by the preconditioned scheme clearly indicates that an extension of the considered explicit time-advancing strategy to more complex test-cases (*e.g.* 3D industrial geometries) could be hardly affordable, due to the high computational cost of the simulation.[25] This result is not restricted to the very simple time-advancing strategy considered in (3.113); indeed, a similar time-step reduction can be observed, for instance, when adopting a classical 4-th-order Runge-Kutta scheme [91].

3.5. Linearized implicit time-advancing

An approximate linearization of the type of (3.13) is proposed in Section 3.5.1, which can be applied to generic Roe numerical flux functions and in particular to (3.64)-(3.68). Then, in Section 3.5.2 the proposed linearization is generalized so as to be applied to preconditioned Roe numerical flux functions of the type of (3.100)-(3.101). Furthermore, a second-order accurate scheme is briefly introduced in Section 3.5.3, defined through a "Defect Correction" technique based on the proposed linearization. Finally, in Sections 3.5.4 and 3.5.5, the discrete solution obtained by exploiting the proposed linearization strategy is respectively validated against a nearly-exact and an exact benchmark.

3.5.1. A linearization of a generic Roe numerical flux function

Linearization of a Roe numerical flux $\phi_{LR}^{(g)\text{ROE}}$

Let $\phi_{LR}^{(g)\text{ROE}}$ denote a Roe numerical flux of the type of (3.39)-(3.42), associated with a generic hyperbolic problem hereafter reminded by superscript (g) (where appropriate).

Note 3.5.1. No specific assumptions on the considered state law are introduced at this stage of the discussion. In particular, the application of the linearization strategy proposed below is not restricted to problems associated with a (generic) barotropic state law.

In order to derive an approximate linearization of $\phi_{LR}^{(g)\text{ROE}}$ of the type of (3.13), two matrices $\mathbf{A}_{LR}^{(g)n}$ and $\mathbf{B}_{LR}^{n}$ are sought, such that the following

---

[25] For instance, the test-case ER-PREC approximately requires 12 hours (CPU time on a laptop with Intel P4 CPU 2.66GHz, 512kB L2 cache and 512MB RAM) for reaching the steady-state.

relation holds:

$$\delta^n \boldsymbol{\phi}_{LR}^{(g)\text{ROE}} \approx \mathbf{A}_{LR}^{(g)n} \cdot \delta^n \mathbf{q}_L^{(g)} + \mathbf{B}_{LR}^{(g)n} \cdot \delta^n \mathbf{q}_R^{(g)} \qquad (3.119)$$

where the increment $\delta^n(\cdot)$ is defined in (3.11). Clearly, the Roe numerical flux is not differentiable and therefore the linearization (3.119) cannot be obtained by a first-order Taylor expansion. Moreover, if the analytical flux $\mathbf{f}^{(g)}$ is a first-order homogeneous function,[26] the following relation is satisfied (by definition):

$$\mathbf{f}^{(g)} = \mathbf{J}^{(g)} \cdot \mathbf{q}^{(g)} \qquad (3.120)$$

where, of course:

$$\mathbf{J}^{(g)} := \partial_{\mathbf{q}^{(g)}} \mathbf{f}^{(g)} \qquad (3.121)$$

and it is possible to recast $\boldsymbol{\phi}_{LR}^{(g)}$ as follows:

$$\boldsymbol{\phi}_{LR}^{(g)} = \mathbf{F} \cdot \mathbf{q}_L^{(g)} + \mathbf{G} \cdot \mathbf{q}_R^{(g)} \qquad (3.122)$$

where $\mathbf{F} = \mathbf{F}(\mathbf{q}_L^{(g)}, \mathbf{q}_R^{(g)})$ and $\mathbf{G} = \mathbf{G}(\mathbf{q}_L^{(g)}, \mathbf{q}_R^{(g)})$ are suitably defined matrices (see below). Then, by assuming $\mathbf{F}$ and $\mathbf{G}$ to be weakly dependent on their arguments, it is possible to choose $\mathbf{A}_{LR}^{(g)n}$ and $\mathbf{B}_{LR}^{(g)n}$ in (3.119) as follows:

$$\mathbf{A}_{LR}^{(g)n} = \mathbf{F}\left(\mathbf{q}_L^{(g)n}, \mathbf{q}_R^{(g)n}\right), \quad \mathbf{B}_{LR}^{(g)n} = \mathbf{G}\left(\mathbf{q}_L^{(g)n}, \mathbf{q}_R^{(g)n}\right).$$

This is a rather classical approach for obtaining an approximate linearization of the type of (3.119) (see *e.g.* [36]). However, as pointed out in [91], there is no uniqueness as far as the choice of $\mathbf{F}$ and $\mathbf{G}$ is concerned. Indeed, by substituting (3.120) into the equalities which are obtained by formally replacing the superscript (A) with (g) in (3.43), it is straightforward to identify the following choices for $\mathbf{F}$ and $\mathbf{G}$:

$$\begin{cases} \mathbf{F} = \mathbf{F}^{(1)}\left(\mathbf{q}_L^{(g)}, \mathbf{q}_R^{(g)}\right) := \mathbf{J}_L^{(g)} - \left(\tilde{\mathbf{J}}_{LR}^{(g)}\right)^{-} \\ \\ \mathbf{G} = \mathbf{G}^{(1)}\left(\mathbf{q}_L^{(g)}, \mathbf{q}_R^{(g)}\right) := \left(\tilde{\mathbf{J}}_{LR}^{(g)}\right)^{-} \end{cases} \qquad (3.123)$$

$$\begin{cases} \mathbf{F} = \mathbf{F}^{(2)}\left(\mathbf{q}_L^{(g)}, \mathbf{q}_R^{(g)}\right) := \left(\tilde{\mathbf{J}}_{LR}^{(g)}\right)^{+} \\ \\ \mathbf{G} = \mathbf{G}^{(2)}\left(\mathbf{q}_L^{(g)}, \mathbf{q}_R^{(g)}\right) := \mathbf{J}_R^{(g)} - \left(\tilde{\mathbf{J}}_{LR}^{(g)}\right)^{+} \end{cases} \qquad (3.124)$$

---

[26] As, for instance, for the Euler equations associated with a perfect gas state law.

where $\mathbf{J}_s^{(g)}$ ($s \in \{L, R\}$) is naturally understood as $\mathbf{J}^{(g)}\left(\mathbf{q}_s^{(g)}\right)$ and, of course, $\tilde{\mathbf{J}}_{LR}^{(g)}$ denotes the relevant Roe matrix. Then, by exploiting (3.123) and (3.124), the following class of approximate linearizations can be introduced [91]:

$$(\mathbf{F}, \mathbf{G}) = \gamma \left(\mathbf{F}^{(1)}, \mathbf{G}^{(1)}\right) + (1 - \gamma) \left(\mathbf{F}^{(2)}, \mathbf{G}^{(2)}\right) \tag{3.125}$$

where γ is a free parameter.

It is possible to propose a linearization of the type of (3.119) even when the first-order homogeneity condition (3.120) is not verified (as, for instance, for the case of the state vectors introduced in Section 2.2, associated with a generic barotropic state law), by virtue of the following:

Proposition 3.5.2. *The Roe numerical flux function* $\boldsymbol{\phi}_{LR}^{(g)\text{ROE}}$ *satisfies the following relation:*

$$\delta^n \boldsymbol{\phi}_{LR}^{(g)\text{ROE}} = \left(\tilde{\mathbf{J}}_{LR}^{(g)n}\right)^+ \cdot \delta^n \mathbf{q}_L^{(g)} + \left(\tilde{\mathbf{J}}_{LR}^{(g)n}\right)^- \cdot \delta^n \mathbf{q}_R^{(g)} + \frac{1}{2} \mathbf{r}_{LR}^{(g),n,n+1} \tag{3.126}$$

where $\mathbf{r}_{LR}^{(g),n,n+1}$ *is defined as follows:*

$$
\begin{aligned}
\mathbf{r}_{LR}^{(g),n,n+1} := {} & \left(\Delta_L \left(\tilde{\mathbf{J}}_{LR}^{(g)}\right)^+ + \Delta_R \left(\tilde{\mathbf{J}}_{LR}^{(g)}\right)^+\right) \cdot \delta^n \mathbf{q}_L^{(g)} \\
& + \left(\Delta_L \left(\tilde{\mathbf{J}}_{LR}^{(g)}\right)^- + \Delta_R \left(\tilde{\mathbf{J}}_{LR}^{(g)}\right)^-\right) \cdot \delta^n \mathbf{q}_R^{(g)} \\
& + \left(\Delta_R \left(\tilde{\mathbf{J}}_{LR}^{(g)}\right)^- - \Delta_L \left(\tilde{\mathbf{J}}_{LR}^{(g)}\right)^+\right) \cdot \Delta^{LR} \mathbf{q}^{(g)n} \\
& + \left(\bar{\Delta}_L \left(\tilde{\mathbf{J}}_{LR}^{(g)}\right)^+ - \bar{\Delta}_R \left(\tilde{\mathbf{J}}_{LR}^{(g)}\right)^-\right) \cdot \Delta^{LR} \mathbf{q}^{(g)n+1}
\end{aligned} \tag{3.127}
$$

with $\Delta^{LR} (\cdot)$ *introduced in (3.37) and:*

$$
\begin{cases}
\Delta_L \left(\tilde{\mathbf{J}}_{LR}^{(g)}\right)^\pm := \tilde{\mathbf{J}}^\pm \left(\mathbf{q}_L^{(g)n+1}, \mathbf{q}_R^{(g)n}\right) - \tilde{\mathbf{J}}^\pm \left(\mathbf{q}_L^{(g)n}, \mathbf{q}_R^{(g)n}\right) \\
\Delta_R \left(\tilde{\mathbf{J}}_{LR}^{(g)}\right)^\pm := \tilde{\mathbf{J}}^\pm \left(\mathbf{q}_L^{(g)n}, \mathbf{q}_R^{(g)n+1}\right) - \tilde{\mathbf{J}}^\pm \left(\mathbf{q}_L^{(g)n}, \mathbf{q}_R^{(g)n}\right) \\
\bar{\Delta}_L \left(\tilde{\mathbf{J}}_{LR}^{(g)}\right)^\pm := \tilde{\mathbf{J}}^\pm \left(\mathbf{q}_L^{(g)n}, \mathbf{q}_R^{(g)n+1}\right) - \tilde{\mathbf{J}}^\pm \left(\mathbf{q}_L^{(g)n+1}, \mathbf{q}_R^{(g)n+1}\right) \\
\bar{\Delta}_R \left(\tilde{\mathbf{J}}_{LR}^{(g)}\right)^\pm := \tilde{\mathbf{J}}^\pm \left(\mathbf{q}_L^{(g)n+1}, \mathbf{q}_R^{(g)n}\right) - \tilde{\mathbf{J}}^\pm \left(\mathbf{q}_L^{(g)n+1}, \mathbf{q}_R^{(g)n+1}\right)
\end{cases} \tag{3.128}
$$

where, finally, $\tilde{\mathbf{J}}^\pm(\mathbf{q}_L^{(g)}, \mathbf{q}_R^{(g)})$ *is obtained by applying the operators defined in (2.2) to the considered Roe matrix* $\tilde{\mathbf{J}}_{LR}^{(g)} = \tilde{\mathbf{J}}(\mathbf{q}_L^{(g)}, \mathbf{q}_R^{(g)})$.

Proof. The proof is reported in Section A.7, for ease of presentation. □

In view of the aforementioned proposition, it is possible to state that, if for all $\left(\mathbf{q}_L^{(g)n}, \mathbf{q}_R^{(g)n}, \mathbf{q}_L^{(g)n+1}, \mathbf{q}_R^{(g)n+1} \right)$ in a same neighbourhood:

$$\left\| \mathbf{r}_{LR}^{(g),n,n+1} \right\| \ll \left\| \left(\tilde{\mathbf{J}}_{LR}^{(g)n} \right)^+ \cdot \delta^n \mathbf{q}_L^{(g)} + \left(\tilde{\mathbf{J}}_{LR}^{(g)n} \right)^- \cdot \delta^n \mathbf{q}_R^{(g)} \right\| \tag{3.129}$$

then, the following approximation can be considered:

$$\delta^n \boldsymbol{\phi}_{LR}^{(g)\text{ROE}} \approx \left(\tilde{\mathbf{J}}_{LR}^{(g)n} \right)^+ \cdot \delta^n \mathbf{q}_L^{(g)} + \left(\tilde{\mathbf{J}}_{LR}^{(g)n} \right)^- \cdot \delta^n \mathbf{q}_R^{(g)} \tag{3.130}$$

It is worth emphasizing that, since the relation (3.126) is obtained by only exploiting the algebraic properties of the Roe numerical flux function (see Section A.7), the linearization (3.130) is independent of the specific state law. Hence, it can be applied to a variety of problems.

Note 3.5.3. Let μ and τ denote the characteristic sizes of the space and time discretizations, respectively. If a certain degree of regularity is assumed for the discrete solution, then $\delta^n \mathbf{q}_s^{(g)}$, $s \in \{L, R\}$, is of the order of τ while $\Delta^{LR} \mathbf{q}^{(g)n}$ is of the order of μ. Furthermore, if the matrices $(\tilde{\mathbf{J}}_{LR}^{(g)})^\pm$ are functions regular enough (*e.g.* Lipschitzian) with respect to their arguments, then the entities in (3.128) are of the order of τ. Hence, the condition (3.129) is verified since $\|\mathbf{r}_{LR}^{(g),n,n+1}\| = O(\tau^2, \tau\mu)$ while the right-hand side term of (3.129) is of the order of τ. In this spirit, the proposed linearization (3.130) is thought to introduce an error which is formally of the order of $O(\tau^2, \tau\mu)$.

Linearization of a Roe numerical flux $\boldsymbol{\phi}_{ij}^{(A)\text{ROE}}$

As far as the linearization of the augmented-1D system of interest is concerned, it is possible to directly exploit the approximate linearization (3.130) at the only cost of formally replacing (g) with (A). Then, by recalling the considerations introduced when deriving $\boldsymbol{\phi}_{ij}^{(A)\text{ROE}}$ from $\boldsymbol{\phi}_{LR}^{(A)\text{ROE}}$ in Section 3.3.1, it is straightforward to generalize the proposed linearization so as to take the orientation of $\hat{\mathbf{v}}_{ij}$ into account. To the purpose, it suffices to choose the right-hand terms of (3.14) as follows:

$$\begin{cases} \mathbf{A}^{(A)} \left(\mathbf{q}_i^{(A)n}, \mathbf{q}_j^{(A)n}, \hat{\mathbf{v}}_{ij} \right) = \left(s_{ij} \, \tilde{\mathbf{J}}_{ij}^{(A)n} \right)^+ \\[4mm] \mathbf{B}^{(A)} \left(\mathbf{q}_i^{(A)n}, \mathbf{q}_j^{(A)n}, \hat{\mathbf{v}}_{ij} \right) = \left(s_{ij} \, \tilde{\mathbf{J}}_{ij}^{(A)n} \right)^- . \end{cases} \tag{3.131}$$

In consideration of (3.131), the linear system (3.15) which is associated
with the proposed linearized implicit scheme reads:

$$\mathbf{M}_{(-1)}^{i,n} \cdot \delta^n \mathbf{q}_{i-1}^{(A)} + \mathbf{M}_{(0)}^{i,n} \cdot \delta^n \mathbf{q}_i^{(A)} + \mathbf{M}_{(+1)}^{i,n} \cdot \delta^n \mathbf{q}_{i+1}^{(A)} = \mathbf{m}^{i,n} \ , \ i \in \mathcal{I} \ (3.132)$$

where:

$$\begin{cases} \mathbf{M}_{(-1)}^{i,n} := \left(s_{i(i-1)} \, \tilde{\mathbf{J}}_{i(i-1)}^{(A)n} \right)^- \\[2mm] \mathbf{M}_{(0)}^{i,n} := \dfrac{\mu_i}{\delta^n t} \mathbf{I} + \left(s_{i(i-1)} \, \tilde{\mathbf{J}}_{i(i-1)}^{(A)n} \right)^+ + \left(s_{i(i+1)} \, \tilde{\mathbf{J}}_{i(i+1)}^{(A)n} \right)^+ \\[2mm] \mathbf{M}_{(+1)}^{i,n} := \left(s_{i(i+1)} \, \tilde{\mathbf{J}}_{i(i+1)}^{(A)n} \right)^- \\[2mm] \mathbf{m}^{i,n} := \boldsymbol{\phi}_{(i-1)i}^{(A)\mathrm{ROE}\,n} - \boldsymbol{\phi}_{i(i+1)}^{(A)\mathrm{ROE}\,n} \end{cases} \qquad (3.133)$$

Moreover, by recalling the definition of s_{ij} given in (3.6) as well as the
relation (3.71), it is possible to simplify the representation of the coefficients on the right-hand side of (3.133) as follows:

$$\begin{cases} \mathbf{M}_{(-1)}^{i,n} = - \left(\tilde{\mathbf{J}}_{(i-1)i}^{(A)n} \right)^+ \\[2mm] \mathbf{M}_{(0)}^{i,n} = \dfrac{\mu_i}{\delta^n t} \mathbf{I} - \left(\tilde{\mathbf{J}}_{(i-1)i}^{(A)n} \right)^- + \left(\tilde{\mathbf{J}}_{i(i+1)}^{(A)n} \right)^+ \\[2mm] \mathbf{M}_{(+1)}^{i,n} = \left(\tilde{\mathbf{J}}_{i(i+1)}^{(A)n} \right)^- \\[2mm] \mathbf{m}^{i,n} := \boldsymbol{\phi}_{(i-1)i}^{(A)\mathrm{ROE}\,n} - \boldsymbol{\phi}_{i(i+1)}^{(A)\mathrm{ROE}\,n} \ . \end{cases} \qquad (3.134)$$

3.5.2. Incorporation of the preconditioning strategy

Let $\boldsymbol{\phi}_{ij}^{(g)\mathrm{ROE},p}$ denote a generic Roe numerical flux function, formally
obtained by replacing (A) with (g) in the definition (3.100)-(3.101).[27]
As for the non-preconditioned case, by exploiting the definition (2.2) together with the property (RM3) introduced in Section 3.3.1, it is possible
to recast $\boldsymbol{\phi}_{ij}^{(g)\mathrm{ROE},p}$ as follows (the superscript (g) is correctly introduced
for the preconditioner as well):

$$\begin{cases} \boldsymbol{\phi}_{ij}^{(g)\mathrm{ROE},p} = s_{ij} \, \mathbf{f}_i^{(g)} + \left(\mathbf{P}_{ij}^{(g)} \right)^{-1} \cdot \left(\mathbf{P}_{ij}^{(g)} \cdot \left(s_{ij} \, \tilde{\mathbf{J}}_{ij}^{(g)} \right) \right)^- \cdot \Delta^{ij} \mathbf{q}^{(g)} \\[2mm] \boldsymbol{\phi}_{ij}^{(g)\mathrm{ROE},p} = s_{ij} \, \mathbf{f}_j^{(g)} - \left(\mathbf{P}_{ij}^{(g)} \right)^{-1} \cdot \left(\mathbf{P}_{ij}^{(g)} \cdot \left(s_{ij} \, \tilde{\mathbf{J}}_{ij}^{(g)} \right) \right)^+ \cdot \Delta^{ij} \mathbf{q}^{(g)} \ . \end{cases} \qquad (3.135)$$

---

[27] A generic state law is assumed at this stage of the discussion. Hence, for instance, besides the
barotropic case specifically treated in the present document it is possible to consider the preconditioned numerical flux discussed in [42].

There is a close formal analogy between the relation (3.135) and the relation (A.49) introduced in Section A.7 for proving the Proposition 3.5.2. In view of this point, it is possible to extend the relevant passages reported in the aforementioned section to the considered preconditioned numerical flux, thus obtaining the following relation:

$$
\begin{cases}
\mathbf{A}^{(g)}\left(\mathbf{q}_i^{(g)n}, \mathbf{q}_j^{(g)n}, \hat{\mathbf{v}}_{ij}\right) = \left(\mathbf{P}_{ij}^{(g)n}\right)^{-1} \cdot \left(\mathbf{P}_{ij}^{(g)n} \cdot \left(s_{ij}\, \tilde{\mathbf{J}}_{ij}^{(g)n}\right)\right)^{+} \\[2mm]
\mathbf{B}^{(g)}\left(\mathbf{q}_i^{(g)n}, \mathbf{q}_j^{(g)n}, \hat{\mathbf{v}}_{ij}\right) = \left(\mathbf{P}_{ij}^{(g)n}\right)^{-1} \cdot \left(\mathbf{P}_{ij}^{(g)n} \cdot \left(s_{ij}\, \tilde{\mathbf{J}}_{ij}^{(g)n}\right)\right)^{-}.
\end{cases}
\tag{3.136}
$$

As for the non-preconditioned case, the proposed linearization (3.136) is only based on the algebraic properties of the Roe numerical flux function and therefore it can be applied to a variety of problems.

The formulation corresponding to the augmented-1D problem considered in the present document is straightforwardly obtained from (3.136) by a trivial change of notation (superscript (A) in place of (g)) and it is reported below for the sake of completeness:

$$
\begin{cases}
\mathbf{A}^{(A)}\left(\mathbf{q}_i^{(A)n}, \mathbf{q}_j^{(A)n}, \hat{\mathbf{v}}_{ij}\right) = \left(\mathbf{P}_{ij}^{(A)n}\right)^{-1} \cdot \left(\mathbf{P}_{ij}^{(A)n} \cdot \left(s_{ij}\, \tilde{\mathbf{J}}_{ij}^{(A)n}\right)\right)^{+} \\[2mm]
\mathbf{B}^{(A)}\left(\mathbf{q}_i^{(A)n}, \mathbf{q}_j^{(A)n}, \hat{\mathbf{v}}_{ij}\right) = \left(\mathbf{P}_{ij}^{(A)n}\right)^{-1} \cdot \left(\mathbf{P}_{ij}^{(A)n} \cdot \left(s_{ij}\, \tilde{\mathbf{J}}_{ij}^{(A)n}\right)\right)^{-}.
\end{cases}
\tag{3.137}
$$

In consideration of the relation (3.137) (which clearly generalizes (3.131)), the linear system (3.15) associated with the proposed preconditioned linearized implicit scheme reads:

$$
\mathbf{L}_{(-1)}^{i,n} \cdot \delta^n \mathbf{q}_{i-1}^{(A)} + \mathbf{L}_{(0)}^{i,n} \cdot \delta^n \mathbf{q}_i^{(A)} + \mathbf{L}_{(+1)}^{i,n} \cdot \delta^n \mathbf{q}_{i+1}^{(A)} = \mathbf{l}^{i,n}, \quad i \in \mathcal{I}
\tag{3.138}
$$

where the relevant coefficients are straightforward generalizations of those reported in (3.134), namely:

$$
\begin{cases}
\mathbf{L}_{(-1)}^{i,n} := -\left(\mathbf{P}_{(i-1)i}^{(A)n}\right)^{-1} \cdot \left(\mathbf{P}_{(i-1)i}^{(A)n} \cdot \tilde{\mathbf{J}}_{(i-1)i}^{(A)n}\right)^{+} \\[3mm]
\mathbf{L}_{(0)}^{i,n} := \dfrac{\mu_i}{\delta^n t}\mathbf{I} - \left(\mathbf{P}_{(i-1)i}^{(A)n}\right)^{-1} \cdot \left(\mathbf{P}_{(i-1)i}^{(A)n} \cdot \tilde{\mathbf{J}}_{(i-1)i}^{(A)n}\right)^{-} \\[3mm]
\qquad + \left(\mathbf{P}_{i(i+1)}^{(A)n}\right)^{-1} \cdot \left(\mathbf{P}_{i(i+1)}^{(A)n} \cdot \tilde{\mathbf{J}}_{i(i+1)}^{(A)n}\right)^{+} \\[3mm]
\mathbf{L}_{(+1)}^{i,n} := \left(\mathbf{P}_{i(i+1)}^{(A)n}\right)^{-1} \cdot \left(\mathbf{P}_{i(i+1)}^{(A)n} \cdot \tilde{\mathbf{J}}_{i(i+1)}^{(A)n}\right)^{-} \\[3mm]
\mathbf{l}^{i,n} := \boldsymbol{\phi}_{(i-1)i}^{(A)\mathrm{ROE},p\,n} - \boldsymbol{\phi}_{i(i+1)}^{(A)\mathrm{ROE},p\,n}.
\end{cases}
$$

3.5.3. A second-order defect-correction scheme

A generalization of the considered linearized implicit scheme, *i.e.* (3.15) coupled with (3.131) or (3.137), is concisely introduced in the present section, based on the "Defect Correction" technique [67] (hereafter DeC) mentioned in the relevant paragraph of Section 3.1.2.

The discrete scheme at hand can be regarded to as an instance of the iterative scheme (3.24). More in detail (of course, the discrete solution here is $\mathbf{z}_h = \mathbf{q}_h^{(A)}$):

- a single iteration is considered: $\lambda_{\max}^n = 1$;
- the time discretization is obtained from the expression (3.17) by considering:
$$k = 1 \quad , \quad \alpha_1 = 1 \, , \, \mathbf{z}_h^{(n,k)} = \mathbf{q}_h^{(A)n}$$
This approximation is first-order accurate and the corresponding truncation error is formally $O(\tau)$, where τ represents the characteristic size of the time discretization;
- the spatial component $\boldsymbol{\psi}_h^{(p)}$ of the non-linear operator $\boldsymbol{\mu}_h^{(p,k)}$ defined in (3.19) is based on the Roe numerical flux. This leads to a first-order accurate discretization (see *e.g.* [39]): $p = 1$. The corresponding error is formally $O(\mu)$, μ representing the characteristic size of the space discretization;
- the linear operator $\mathbf{J}_h^{(q,k)}$ is defined by choosing $q = 1$. The term $\delta\boldsymbol{\psi}_h^{(q)}$ appearing in the relevant definition (3.21) is constructed, in particular, by exploiting the approximate linearization (3.130) which, as pointed out in Note 3.5.3 (Section 3.5.1), formally introduces a discretization error $O(\tau^2, \tau\mu)$.

The resulting scheme clearly introduces a discretization error $O(\mu, \tau)$ and therefore it is only first-order accurate.

It is possible to increase the accuracy of the aforementioned scheme up to the second order by adopting a DeC strategy, as briefly outlined in Section 3.1.2. To the purpose:

- a second-order backward finite difference approximation is derived from (3.17) by means of the following settings:
$$k = 2 \quad , \quad \alpha_2 = \frac{1 + 2\theta}{1 + \theta} \, , \, \mathbf{z}_h^{(n,k)} = (1 + \theta) \, \mathbf{q}_h^{(A)n} - \frac{\theta^2}{1 + \theta} \mathbf{q}_h^{(A)n-1}$$
where:
$$\theta := \frac{\delta^n t}{\delta^{n-1} t} .$$
The truncation error associated with the approximation at hand is formally $O(\tau^2)$;

- a second-order spatial discretization $\boldsymbol{\psi}_h^{(p)}$, with $p = 2$, is introduced by performing a MUSCL reconstruction [106, 107, 108] before evaluating the Roe numerical flux. According to this strategy, the Roe numerical flux between the cells C_i and C_j (towards C_j) is computed as follows:

$$\boldsymbol{\phi}^{(A)\text{ROE}}\left(\mathbf{q}_{[i]j}^{(A)}, \mathbf{q}_{i[j]}^{(A)}, \hat{\boldsymbol{v}}_{ij}\right)$$

where $\boldsymbol{\phi}^{(A)\text{ROE}}$ represents the usual Roe flux function[28] while $\mathbf{q}_{[i]j}^{(A)}$ and $\mathbf{q}_{i[j]}^{(A)}$ denote suitably extrapolated values at the interface between C_i and C_j, respectively on the side of C_i and C_j. The considered extrapolation is constrained (by exploiting the starting, piece-wise constant, discrete solution) in a non-linear fashion, so as to avoid spurious oscillations (see $e.g.$ [39, 64, 98] and many references cited therein);

- in the spirit of the DeC approach [67], a value $q < p$ is chosen for containing the computational cost associated with the inversion of the linear operator $\mathbf{J}_h^{(q,k)}$ (see Section 3.1.2). Hence, in particular, the value $q = 1$ is adopted, as for the case discussed in the corresponding point of the previous list. In other words, the term $\delta \boldsymbol{\psi}_h^{(q)}$ is constructed by applying the proposed approximate linearization (3.130) to the starting, piece-wise constant, numerical solution. Of course, the corresponding discretization error is still $O(\tau^2, \tau\mu)$.

In view of the aforementioned points, a single iteration ($i.e.$ $\lambda_{\text{max}}^n = 1$) of the scheme (3.24) yields a discretization error $O(\mu^2, \tau^2, \tau\mu) = O(\epsilon^2)$, with $\epsilon := \max(\mu, \tau)$ and therefore a second-order accurate solution is obtained. Moreover, on the basis of some preliminary carried out numerical experiments [90], a sensible improvement in the solution behaviour is observed by slightly increasing λ_{max}^n, $e.g.$ by performing 2 or 3 iterations. Hence, the DeC seems to be a promising strategy for defining high-order, efficient schemes based on the proposed linearization (3.130); further investigations on this subject is definitely recommended.

3.5.4. Numerical results for smooth flows

Benchmarks

The benchmark already introduced in Section 3.4.4 (namely the quasi-1D flow within a duct having variable cross-sectional area) and summarized, in particular, in Table 3.6 is considered here, with the aim of directly comparing the proposed linearized implicit time-advancing strategy with

---

[28] Any additional superscript, like that one denoting the preconditioning technique discussed in the previous sections, is here dropped, for the sake of simplicity.

the explicit one given by (3.113). The relevant definitions/considerations are tacitly recalled from the aforementioned section.

Discrete scheme, initial and boundary conditions

A linearized implicit discrete scheme can be derived from (3.112) by following the procedure which permits to obtain (3.15) from (3.9), at the only cost of extending the linearization to the term $\mathbf{s}^{(x)}$ as follows:

$$\mathbf{s}_i^{(x)n+1} \approx \mathbf{s}_i^{(x)n} + \mathbf{S}_i^{(x)n} \cdot \delta^n \mathbf{q}_i^{(x)} \tag{3.139}$$

where:

$$\mathbf{S}_i^{(x)n} := \partial_{\mathbf{q}^{(x)}} \partial \mathbf{s}^{(x)} \left(\mathbf{q}_i^{(x)n} \right) = \begin{pmatrix} 0 & 1 \\ -\left(u_i^n\right)^2 & 2u_i^n \end{pmatrix}$$

In particular, it suffices to respectively incorporate $\mathbf{s}_i^{(x)n}$ and $\mathbf{S}_i^{(x)n}$ into the right-hand side term and the diagonal coefficient of a linear system of the type of (3.138), namely:

$$\hat{\mathbf{L}}_{(-1)}^{i,n} \cdot \delta^n \mathbf{q}_{i-1}^{(x)} + \hat{\mathbf{L}}_{(0)}^{i,n} \cdot \delta^n \mathbf{q}_i^{(x)} + \hat{\mathbf{L}}_{(+1)}^{i,n} \cdot \delta^n \mathbf{q}_{i+1}^{(x)} = \hat{\mathbf{I}}^{i,n} , \quad i \in \mathcal{I} \tag{3.140}$$

with:

$$\begin{cases} \hat{\mathbf{L}}_{(-1)}^{i,n} := - \left(\mathbf{P}_{(i-1)i}^{(x)n} \right)^{-1} \cdot \left(\mathbf{P}_{(i-1)i}^{(x)n} \cdot \tilde{\mathbf{J}}_{(i-1)i}^{(x)n} \right)^+ \\[2mm] \hat{\mathbf{L}}_{(0)}^{i,n} := \dfrac{\mu_i}{\delta^n t} \mathbf{I} - \mathbf{S}_i^{(x)n} - \left(\mathbf{P}_{(i-1)i}^{(x)n} \right)^{-1} \cdot \left(\mathbf{P}_{(i-1)i}^{(x)n} \cdot \tilde{\mathbf{J}}_{(i-1)i}^{(x)n} \right)^- \\[2mm] \qquad + \left(\mathbf{P}_{i(i+1)}^{(x)n} \right)^{-1} \cdot \left(\mathbf{P}_{i(i+1)}^{(x)n} \cdot \tilde{\mathbf{J}}_{i(i+1)}^{(x)n} \right)^+ \\[2mm] \hat{\mathbf{L}}_{(+1)}^{i,n} := \left(\mathbf{P}_{i(i+1)}^{(x)n} \right)^{-1} \cdot \left(\mathbf{P}_{i(i+1)}^{(x)n} \cdot \tilde{\mathbf{J}}_{i(i+1)}^{(x)n} \right)^- \\[2mm] \hat{\mathbf{I}}^{i,n} := \mathbf{s}_i^{(x)n} + \boldsymbol{\phi}_{(i-1)i}^{(x)\text{ROE},p\,n} - \boldsymbol{\phi}_{i(i+1)}^{(x)\text{ROE},p\,n} . \end{cases}$$

The uniform flow field given by (3.114) is chosen as IC while, as far as the BCs are concerned:

- the Dirichlet-like BC (3.116) clearly implies that:

$$\delta^n \mathbf{q}_0^{(x)} = 0$$

and therefore it is naturally implemented as follows:

$$\hat{\mathbf{L}}_{(0)}^{1,n} \cdot \delta^n \mathbf{q}_1^{(x)} + \hat{\mathbf{L}}_{(+1)}^{1,n} \cdot \delta^n \mathbf{q}_2^{(x)} = \hat{\mathbf{I}}^{1,n}$$

- the transmissive BC (3.117) clearly implies that:

$$\delta^n \mathbf{q}_{N_c+1}^{(x)} = \delta^n \mathbf{q}_{N_c}^{(x)}$$

and therefore it is naturally implemented as follows:

$$\hat{\mathbf{L}}_{(-1)}^{N_c,n} \cdot \delta^n \mathbf{q}_{N_c-1}^{(x)} + \left(\hat{\mathbf{L}}_{(0)}^{N_c,n} + \hat{\mathbf{L}}_{(+1)}^{N_c,n} \right) \cdot \delta^n \mathbf{q}_{N_c}^{(x)} = \hat{\mathbf{i}}^{N_c,n} \tag{3.141}$$

Test-cases

The considered test-cases are summarized in Table 3.8, in which BN denotes the considered benchmark (described in Table 3.6, Section 3.4.4).

Table 3.8. Considered test-cases for the discrete scheme (3.140), based on the preconditioned numerical flux (3.88)-(3.90).

Test-case	Benchmark	μ	β^2	τ
IR-NOPREC	BN	10	1	$\approx \infty$
IR-PREC	BN	10	10^{-6}	$\approx \infty$

In particular:

- the x-domain $[x_{min}, x_{max}]$ (with x_{min} and x_{max} defined in the aforementioned Table 3.6) is uniformly discretized (with cells having size μ) for both cases;
- for the test-case IR-NOPREC the proposed preconditioning strategy is not activated (indeed for $\beta^2 = 1$ the preconditioner (3.93) reduces to the identity matrix and therefore it does not modify the numerical flux function). Conversely, the test-case IR-PREC exploits the preconditioning strategy at hand, by choosing β_{ref} in (3.97) exactly equal to 1;
- it turns out that, for both the considered test-cases, a practically "unbounded" time-step can be adopted for advancing the numerical solution by means of the proposed linearized implicit strategy. In the carried out numerical experiments τ has been increased up to 10^5, thus reaching the steady-state solution in a very few (namely 2 to 5) iterations; the required CPU time is practically negligible.

The corresponding numerical solutions, shown in Figures 3.24 and 3.25, are indistinguishable from their counterparts obtained by the explicit time-advancing (reported in Figures 3.22 and 3.23, Section 3.4.4).

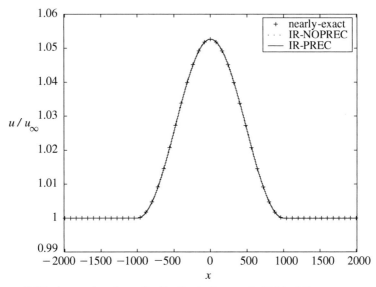

Figure 3.24. Approximation of u for the test-cases in Table 3.8.

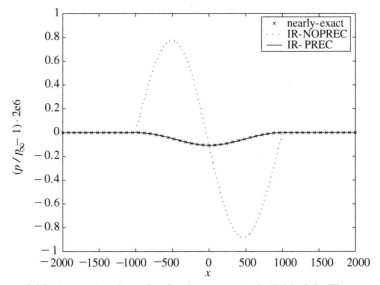

Figure 3.25. Approximation of p for the test-cases in Table 3.8. The pressure variation on the y-axis is scaled for ease of readability.

A local preconditioning strategy

It may be worth investigating the behaviour of the preconditioned numerical flux as the local Mach number of the flow field undergoes non-negligible variations. Indeed, under this circumstance, it may be difficult to identify a unique Mach number $M_\star$ which is representative of the entire flow field and, consequently, the definition of the preconditioning parameter β^2 in (3.97) may be not straightforward.

To the purpose, it is possible to consider the nozzle flow introduced in Section 3.4.4. More precisely, by varying the parameter σ (which determines the variation of cross-sectional area through (3.110)), it is possible to control the velocity u in the duct as well as the corresponding (local) Mach number, since the sound speed is constant for the considered benchmark ($a = 10^3$). Thus, for instance, by choosing $\sigma = 4.5 \cdot 10^{-1}$ (while keeping the other settings in Table 3.6, Section 3.4.4), the minimum value of α provided by (3.111) is $\alpha_{\min} = 0.1$. The maximum value of u can be obtained from (3.108), since the flow turns out to be nearly-incompressible for the present case as well. In particular, the maximum value of u (taken in correspondence of the minimum cross-sectional area of the duct, hereafter referred to as throat as well) is roughly $10 \cdot u_\infty$. Hence, if M_∞ denotes the inlet Mach number ($M_\infty = 10^{-3}$ for the present case), the Mach number at the throat is $M_{\text{throat}} = 10 \cdot M_\infty$. Then, by choosing M_∞ as representative of the whole flow field: $M_\star = M_\infty$ and by choosing $\beta^2 = M_\star^2 (= 10^{-6})$ as preconditioning parameter, it follows that $\beta^2 = M_{\text{throat}}^3 \neq M_{\text{throat}}^2$. As a consequence, it is reasonable to expect a less accurate numerical solution near the throat. This is confirmed by the curves labelled with "GLOBAL-PREC" in Figures 3.26 and 3.27, which are computed using the settings of the test-case IR-PREC reported in Table 3.8 above. It may be worth noticing that the discrepancy between the numerical and the nearly-exact solution in the aforementioned figures does not appreciably propagate towards the inlet section, since the adopted Dirichlet-like BC does not allow for a substantial variation of the state vector to take place.

In view of the aforementioned considerations, it makes sense to investigate the effects which are produced on the numerical solution by replacing the original preconditioning parameter β^2 with a new one, say β_{ij}^2, taking into account the local Mach number. More in detail, since the preconditioner acts (as the numerical flux, of course) at the cell interface, it seems reasonable to relate β_{ij}^2 to a certain Mach number $\bar{M}_{ij}$ which can be considered representative of both the adjacent state vectors $\mathbf{q}_i^{(x)}$ and $\mathbf{q}_j^{(x)}$. In consideration of the fact that the Roe flux between $\mathbf{q}_i^{(x)}$ and $\mathbf{q}_j^{(x)}$ is essentially based on the averaging defined, as the name suggests, by the Roe averages (see Note 3.3.4 in Section 3.3.1), the following choice seems to be quite natural:

$$\bar{M}_{ij} := \frac{|u_{ij}|}{a_{ij}}.$$

Then, a definition for β_{ij}^2 may be the following:

$$\beta_{ij}^2 := 1 - \exp\left(-\kappa_{ij} \cdot \left(\bar{M}_{ij}\right)^2\right) \tag{3.142}$$

where $\kappa_{ij} = O(1)$ is a free parameter. Indeed, according to the definition above:

- for $\bar{M}_{ij} \to 0$, $\beta_{ij}^2 \to \kappa_{ij} \cdot (\bar{M}_{ij})^2$, somehow (locally) recovering the original relation (3.97);
- as $\bar{M}_{ij}$ increases, $\beta_{ij}^2 \to 1$ and the effects of the preconditioning strategy correctly disappear. The parameter κ_{ij}, in particular, can be modelled in order to control the transition under consideration.

The numerical solution which is obtained by adopting (3.142) with $\kappa_{ij} = 1$ (while keeping the remaining settings of the aforementioned test-case IR-PREC) is shown by the curves labelled with "LOCAL-PREC" in Figures 3.26 and 3.27. The considered solution turns out to be more accurate than that one obtained by the global preconditioning technique, even if there are still discrepancies with respect to the nearly-exact benchmark. In view of this result, it seems reasonable to further investigate (even heuristic) generalizations of the considered preconditioning technique, like (3.142), in order to accurately simulate flow fields characterized by non-negligible variations of the local Mach number. Such a study is postponed to a subsequent research activity.

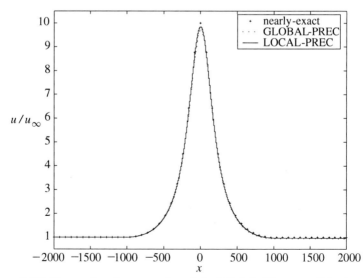

Figure 3.26. Comparison between the original ("global") preconditioning strategy and the modified ("local") one: effects on u.

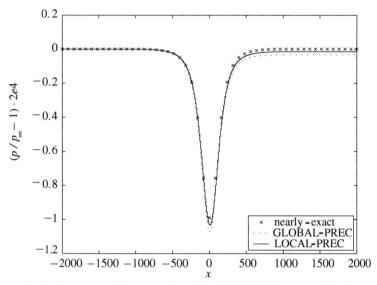

Figure 3.27. Comparison between the original ("global") preconditioning strategy and the modified ("local") one: effects on p. The pressure variation on the y-axis is scaled for ease of readability.

3.5.5. Numerical results for non-smooth flows at low Mach numbers

In order to directly compare the proposed linearized implicit time-advancing strategy with an explicit one, the implicit counterpart of the test-case ER-2-3 described in Table 3.4 (Section 3.3.2) is considered.

The benchmark description as well as any relevant definition is tacitly recalled from the aforementioned section. The discrete scheme (3.132) is considered, associated with a constant time-step τ. The implementation of the assumed BCs (3.28) reads (compare with (3.141)):

$$\begin{cases} \left(\mathbf{M}^{1,n}_{(-1)} + \mathbf{M}^{1,n}_{(0)} \right) \cdot \delta^n \mathbf{q}^{(A)}_1 + \mathbf{M}^{1,n}_{(+1)} \cdot \delta^n \mathbf{q}^{(A)}_2 = \mathbf{m}^{1,n} \\ \mathbf{M}^{N_c,n}_{(-1)} \cdot \delta^n \mathbf{q}^{(A)}_{N_c-1} + \left(\mathbf{M}^{N_c,n}_{(0)} + \mathbf{M}^{N_c,n}_{(+1)} \right) \cdot \delta^n \mathbf{q}^{(A)}_{N_c} = \mathbf{m}^{N_c,n} . \end{cases} \tag{3.143}$$

The considered test-cases are reported in Table 3.9; the behaviour of the corresponding numerical solutions is shown in Figures 3.28-3.31.
Some entities which can be exploited for evaluating the accuracy as well as the computational cost of the considered simulations are reported in Table 3.10 (to be compared with the relevant row of Table 3.5 in Section 3.3.2). It should be noticed that:

- the estimate $\tilde{c}^{(CFL)}$ is directly proportional to τ since its definition (3.33) is based on the largest wave speed of the benchmark RP (which,

Table 3.9. Considered test-cases for the discrete scheme (3.132), based on the numerical flux (3.64)-(3.68).

Test-case	Benchmark	μ	(n_L, n_R)	τ
IR2-3-1	B2	1	$(2, 2) \cdot 10^3$	$5 \cdot 10^{-4}$
IR2-3-2	B2	1	$(2, 2) \cdot 10^3$	$5 \cdot 10^{-3}$
IR2-3-3	B2	1	$(2, 2) \cdot 10^3$	$5 \cdot 10^{-2}$

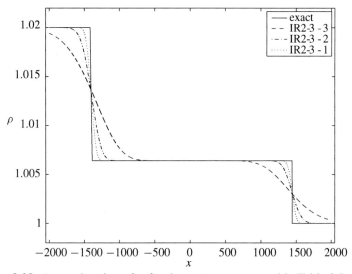

Figure 3.28. Approximation of ρ for the test-cases reported in Table 3.9.

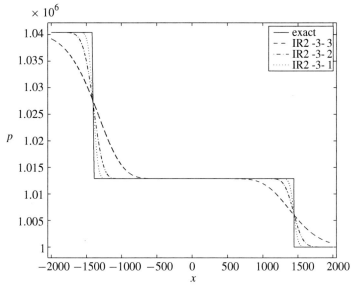

Figure 3.29. Approximation of p for the test-cases reported in Table 3.9.

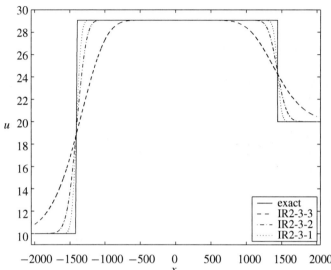

Figure 3.30. Approximation of u for the test-cases reported in Table 3.9.

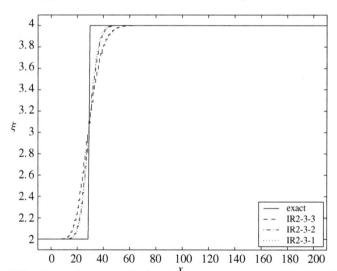

Figure 3.31. Approximation of ξ for the test-cases reported in Table 3.9. The x-range is cut for ease of readability.

Table 3.10. CFL estimate, CPU time and error estimates for the test-cases reported in Table 3.9.

Test-case	$\tilde{c}^{(CFL)}$	t_{CPU}	$e(\rho)$	$e(p)$	$e(u)$	$e(\xi)$
IR2-3-1	7.2e-1	≈ 5 min.	0.0707	0.1419	3.8683	1.0599
IR2-3-2	7.2	≈ 30 sec.	0.1111	0.2231	6.0785	1.0956
IR2-3-3	7.2e1	≈ 4 sec.	0.1992	0.3997	10.8928	1.3401

of course, is not affected by the numerical set-up) for all the considered test-cases. In particular, $\tilde{c}^{(CFL)}$ assumes the same value for the test-cases ER2-3 and IR2-3-1, for which the same time-step is adopted. The linearized implicit scheme does not suffer from the stability restriction encountered in the explicit case (coefficients $\tilde{c}^{(CFL)} >$ 1 can be adopted) and it is therefore more efficient than the explicit one. However, as discussed in more detail in Section 3.5.6, the proposed linearized implicit scheme is not unconditionally stable and the largest time-step which can be adopted seems to be somehow related to the magnitude of the discontinuity introduced by the IC of the underlying benchmark RP;

- the CPU time (on a laptop with Intel P4 CPU 2.66GHz, 512kB L2 cache, 512MB RAM) turns out to be inversely proportional to τ. The considered implementation of the implicit time-advancing turns out to be slower than the explicit one of about one order of magnitude;[29]
- despite small differences (which may be partly addressed to the implementation), the numerical solution obtained by the considered implicit scheme turns out to be as accurate as that one obtained by the considered explicit scheme (compare the test-case IR2-3-1 in Table 3.10 with the test-case ER2-3 in Table 3.5, Section 3.3.2). Moreover, the accuracy of the numerical solution provided by the linearized implicit scheme rapidly degrades as the time-step is increased (for a chosen space discretization), as shown in Figure 3.32.[30] Hence, the largest time-step which can be adopted when using the proposed linearized implicit scheme for unsteady simulations could be bounded by chosen accuracy requirements even before reaching the aforementioned stability limit.

3.5.6. Numerical results for non-smooth flows at generic Mach numbers

Up to the present section, all the considered numerical experiments have focused attention on low Mach number flows, because of widely discussed reasons. However, since the proposed linearized implicit time-advancing can be applied to generic Roe numerical flux functions and

---

[29] This result does not seem to be closely related to the specific implementation of the implicit solver. Indeed, comparable CPU times have been obtained by considering two different solvers for the linear system of interest (namely a library routine for banded matrices and the block version of the Thomas algorithm [79] for tri-diagonal systems).

[30] It should be noticed that, with the only exception of $e(\xi)$ (whose measure is, in general, more susceptible to errors due to the specific shape of the relevant curve), the remaining curves exhibit a similar trend.

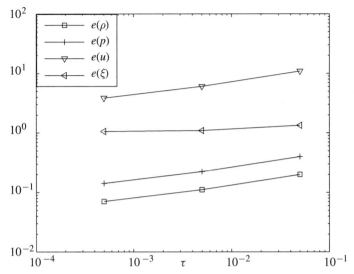

Figure 3.32. Plot of the error estimates for the test-cases reported in Table 3.10.

therefore to a variety of problems, it is of interest to also investigate the behaviour of the discrete solution when the characteristic Mach number of the flow (if any) is not close to zero.

To the purpose, the benchmark summarized in Table 3.11 is considered. In this table, κ, $\varkappa$ and γ refer to the chosen convex state law (2.66), ρ_L, u_L, ξ_L, ρ_R, u_R and ξ_R characterize the initial condition (IC) of a Riemann problem (RP) and t_{eval} denotes the time at which the considered solution is picked.

Table 3.11. Considered benchmark.

Benchmark	κ	$\varkappa$	γ	ρ_L	u_L	ξ_L	ρ_R	u_R	ξ_R	t_{eval}
B3	1	1	0	1	0.9	2	1	−0.9	4	1

The adopted state law introduces a constant sound speed $a = \sqrt{\kappa} = 1$ and therefore the unperturbed "left" and "right" regions of the relevant RP are characterized by a local Mach number equal to 0.9. On the other hand, due to the symmetry of the chosen IC, $u_\star = 0$ (easily derived by averaging the expressions in (2.84) for $\rho_L = \rho_R$) and consequently the local Mach number undergoes a noticeable variation through the domain.[31] In this sense, the present benchmark introduces discontinuities

---

[31] Sonic conditions are deliberately avoided, see Note 3.3.12 in Section 3.3.1.

which are stronger than those considered in previous numerical experiments. Besides the (stationary, since $u_\star = 0$) contact discontinuity, two symmetrical shock waves appear as part of the solution, travelling with speed $\tilde{s} \approx 0.65$.

The discrete scheme (3.132) is considered, associated with a uniform space discretization having measure $\mu = 10^{-2}$ and a constant time-step τ. It may be worth noticing that the adopted space discretization is only apparently finer than that one considered in previous numerical experiments, *e.g.* those reported in Table 3.9. Indeed, a relevant parameter is $\mu/\tilde{\mu}$, where $\tilde{\mu}$ denotes a characteristic length scale of the problem. For the RP at hand, $\tilde{\mu} \approx \tilde{s} \, t_{\text{eval}} = O(1)$ and $\mu/\tilde{\mu} = O(10^{-2})$ while for the test-cases reported in Table 3.9 (for which the largest wave speed is of the order of 10^3) $\mu/\tilde{\mu} = O(10^{-3})$. In a similar manner, a relevant parameter for the time discretization is τ/t_{eval}. The boundary conditions (3.143) are adopted for closing the problem.

The considered test-cases are reported in Table 3.12. The numerical approximation of ρ (*i.e.* p, since $p = \rho$ according to the considered state law) and u are shown in Figures 3.33 and 3.34, respectively. An example of the numerical approximation of ξ is reported in Figure 3.35.

Table 3.12. Considered test-cases.

Test-case	Benchmark	μ	(n_L, n_R)	τ
IR-M09-1	B3	10^{-2}	$(2,2) \cdot 10^2$	$5 \cdot 10^{-3}$
IR-M09-2	B3	10^{-2}	$(2,2) \cdot 10^2$	$1 \cdot 10^{-2}$
IR-M09-3	B3	10^{-2}	$(2,2) \cdot 10^2$	$1 \cdot 10^{-1}$

It is worth remarking that:

- by increasing the time-step (for a fixed space discretization) the numerical scheme becomes unstable. More in detail:
  - for the test-cases IR-M09-1 the coefficient $\tilde{c}^{(\text{CFL})}$ defined in (3.33) is approximately equal to 0.33 and an explicit time-advancing running with the same $\tilde{c}^{(\text{CFL})}$ turns out to be stable as well;
  - for the test-cases IR-M09-2 and IR-M09-3, $\tilde{c}^{(\text{CFL})}$ is approximately equal to 0.65 and 6.5, respectively. By further increasing the time-step ($1.25 \cdot 10^{-1} < \tau < 2.00 \cdot 10^{-1} \Rightarrow \tilde{c}^{(\text{CFL})} \approx 10$), a blow-up occurs after a few iterations: the numerical solution becomes unstable since the beginning of the simulation, in correspondence of the discontinuity associated with the benchmark RP.

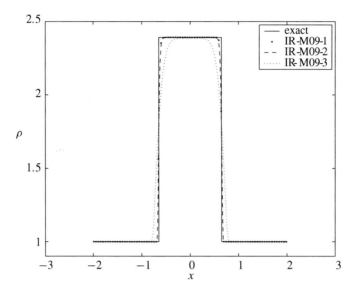

Figure 3.33. Approximation of ρ (*i.e.* p) for the test-cases reported in Table 3.12.

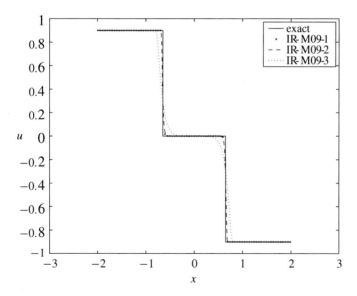

Figure 3.34. Approximation of u for the test-cases reported in Table 3.12.

This observation suggests the existence of a stability limit for the linearized implicit time-advancing. By comparing the maximum $\tilde{c}^{(CFL)}$ which can be adopted in the present case with e.g that one associated

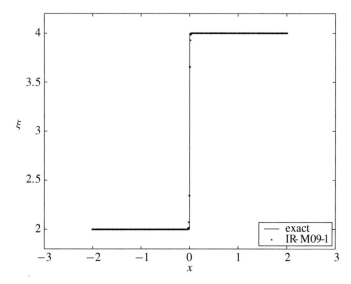

Figure 3.35. Approximation of ξ for the test-case IR-M09-1 reported in Table 3.12.

with the test-case IR2-3-3 in Table 3.10, it is possible to put forward the hypothesis that the stability limit under consideration somehow becomes more severe as the magnitude of some relevant discontinuities associated with the underlying RP increases.[32] For instance, the variation of the density for the benchmark B3 (associated with the test-cases in Table 3.12) is $O(1)$ (see Figure 3.33) while that one for the benchmark B2 (associated with the test-cases in Table 3.10) is $O(10^{-2})$; an even more considerable difference is observed when considering the variation of the Mach number.

The aforementioned hypothesis may be supported by the fact that for smooth solutions this stability problem does not appear (see *e.g.* the nozzle flow discussed in Section 3.5.4). On the other hand, some numerical experiments involving stronger initial discontinuities (not reported here for brevity) exhibit an even narrow stability margin. After all, the presence of a discontinuity within the flow field makes it more difficult to apply the proposed linearization strategy (3.130); in particular, it is likely to violate the condition (3.129) in the neighbourhood of the discontinuity. In summary, as soon as considerable discontinuities appear within the flow field, a time-step reduction could be required in order to keep the proposed linearized implicit time-

[32] The jump of the passive scalar ξ may play a minor role, since it does not directly affect the considered Roe linearization.

advancing algorithm stable,[33] thus reducing the efficiency of the corresponding numerical scheme. However, this stability problem does not appear to be related to the specific linearization which is proposed in the present work; indeed, it affects other linearizations as well, as discussed in Section 3.5.7;

- as for the test-cases presented in Section 3.5.5, the accuracy of the considered numerical solutions rapidly degrades as the time-step is increased and the largest time-step which can be adopted for unsteady simulations could be bounded by chosen accuracy requirements even before reaching the aforementioned stability limit;

- the considered linearized implicit scheme is able to approximate the contact discontinuity with a reasonably good accuracy (*i.e.* a few cells), as shown in Figure 3.35 for the test-case IR-M09-1. This result is not obvious, since in general it is not a trivial task to approximate slowly moving (in particular, stationary) contact discontinuities [98].

3.5.7. A remark on the linearization technique

It seems valuable to address the issue of whether the stability constraint highlighted in Section 3.5.6 is specific to the proposed linearization technique or not. To the purpose, two linearizations of the Roe flux function are recalled, of the type of (3.13). As for the proposed linearization (3.130), the aforementioned ones can be applied to a generic Roe numerical flux function $\phi_{LR}^{(g)\mathrm{ROE}}$.

(L1) By adopting the following approximations:

$$\delta^n \mathbf{f}^{(g)} \approx \mathbf{J}^{(g)n} \cdot \delta^n \mathbf{q}^{(g)} \tag{3.144}$$

$$\delta^n \left| \tilde{\mathbf{J}}_{LR}^{(g)} \right| \approx \mathbf{0} \tag{3.145}$$

where $\mathbf{J}^{(g)}$ denotes the Jacobian defined by (3.121) and $\tilde{\mathbf{J}}_{LR}^{(g)}$ represents the relevant Roe matrix, the variation of the centred and upwind components of $\phi_{LR}^{(g)\mathrm{ROE}}$ can be expressed as follows:

$$\delta^n \phi_{c,LR}^{(g)\mathrm{ROE}} \approx \frac{1}{2} \left(\mathbf{J}_L^{(g)n} \cdot \delta^n \mathbf{q}_L^{(g)} + \mathbf{J}_R^{(g)n} \cdot \delta^n \mathbf{q}_R^{(g)} \right) \tag{3.146}$$

$$\delta^n \phi_{u,LR}^{(g)\mathrm{ROE}} \approx -\frac{1}{2} \left| \tilde{\mathbf{J}}_{LR}^{(g)n} \right| \cdot \left(\delta^n \mathbf{q}_R^{(g)} - \delta^n \mathbf{q}_L^{(g)} \right). \tag{3.147}$$

---

[33] In a practical computational set-up the time-step can be modulated, possibly by an adaptive strategy, so as to mitigate the stability problem under consideration.

Consequently, a linearization of the type of (3.119) can be straight-forwardly introduced, involving the following coefficients:

$$
\begin{cases}
\mathbf{A}_{LR}^n = \dfrac{1}{2} \left(\mathbf{J}_L^{(g)n} + \left| \tilde{\mathbf{J}}_{LR}^{(g)n} \right| \right) \\[4mm]
\mathbf{B}_{LR}^n = \dfrac{1}{2} \left(\mathbf{J}_R^{(g)n} - \left| \tilde{\mathbf{J}}_{LR}^{(g)n} \right| \right).
\end{cases}
\tag{3.148}
$$

The linearization (3.148) is exploited in [31] for defining a linearized implicit time-advancing technique.

(L2) By defining a matrix $\mathbf{J}^{(g)\star} = \mathbf{J}^{(g)\star}\left(\mathbf{q}^{(g)}\right)$ which mimics the first-order homogeneity property (3.120) as follows:

$$
\mathbf{f}^{(g)} = \mathbf{J}^{(g)\star} \cdot \mathbf{q}^{(g)}
\tag{3.149}
$$

it is possible to introduce the following linearization (formally similar to (3.144)):

$$
\delta^n \mathbf{f}^{(g)} \approx \mathbf{J}^{(g)\star n} \cdot \delta^n \mathbf{q}^{(g)}.
$$

Then, it is possible to replace (3.146) with the following expression:

$$
\delta^n \boldsymbol{\phi}_{c,LR}^{(g)\mathrm{ROE}} \approx \dfrac{1}{2} \left(\mathbf{J}_L^{(g)\star n} \cdot \delta^n \mathbf{q}_L^{(g)} + \mathbf{J}_R^{(g)\star n} \cdot \delta^n \mathbf{q}_R^{(g)} \right).
$$

Finally, by keeping the approximation (3.145) (and, consequently, (3.147)) the following additional linearization can be introduced:

$$
\begin{cases}
\mathbf{A}_{LR}^n = \dfrac{1}{2} \left(\mathbf{J}_L^{(g)\star n} + \left| \tilde{\mathbf{J}}_{LR}^{(g)n} \right| \right) \\[4mm]
\mathbf{B}_{LR}^n = \dfrac{1}{2} \left(\mathbf{J}_R^{(g)\star n} - \left| \tilde{\mathbf{J}}_{LR}^{(g)n} \right| \right)
\end{cases}
\tag{3.150}
$$

which, of course, is similar to the previous one (3.148). The definition of the matrix $\mathbf{J}^{(g)\star}$) and, consequently, the linearization (3.150) are introduced in [4] in order to define a linearized implicit time-advancing technique.

Note 3.5.4. In general, the definition of $\mathbf{J}^{(g)\star}$ is not unique, as shown by the following example based on the basic-1D state vector $\mathbf{q}^{(x)}$ (introduced in Section 2.2.3). Let α_{mn}, with $m, n \in \{1, 2\}$, denote the mn-th component of $\mathbf{J}^{(x)\star}$. Then, the following relation must be verified, by the definition (3.149):

$$
\begin{pmatrix} \rho u \\ \rho u^2 + p \end{pmatrix} = \begin{pmatrix} \alpha_{11} & \alpha_{12} \\ \alpha_{21} & \alpha_{22} \end{pmatrix} \cdot \begin{pmatrix} \rho \\ \rho u \end{pmatrix}.
$$

By virtue of the fact that ρ and u are independent of each other, it necessarily follows that $\alpha_{11} = 0$ and $\alpha_{12} = 1$, while the remaining relation leads to the following equation:

$$\alpha_{21} \, \rho + \alpha_{22} \, \rho \, u = \rho \, u^2 + p \qquad (3.151)$$

which admits an infinite number of solutions. By assuming, for instance, $\alpha_{22} = 2 \, u$, the matrix $\mathbf{J}^{(x)\star}$ can be expressed as follows:

$$\mathbf{J}^{(x)\star} = \begin{pmatrix} 0 & 1 \\ a^2 - u^2 + \left(\dfrac{p}{\rho} - a^2 \right) & 2 \, u \end{pmatrix} \qquad (3.152)$$

while, by assuming $\alpha_{21} = a^2 - u^2$, the following representation is obtained:

$$\mathbf{J}^{(x)\star} = \begin{pmatrix} 0 & 1 \\ a^2 - u^2 & 2 \, u + u^{-1} \left(\dfrac{p}{\rho} - a^2 \right) \end{pmatrix}. \qquad (3.153)$$

The matrix (3.152), in particular, is exploited in [4].

It is worth noticing that the aforementioned linearization techniques (L1) and (L2) may coincide with each other. For instance, if the adopted barotropic state law is first-order homogeneous:

$$p = \frac{dp}{d\rho} \, \rho = a^2 \, \rho \qquad (3.154)$$

then the expressions (3.152) and (3.153) become equal to each other and $\mathbf{J}^{(x)\star}$ reduces to the relevant Jacobian $\mathbf{J}^{(x)}$. As a result, the aforementioned linearization techniques (L1) and (L2) coincide with each other.

A few simulations have been carried out, also involving the aforementioned linearizations (L1) and (L2). In particular, the test-case IR-M09-3 reported in Table 3.12 above has been considered. The state law associated with the corresponding benchmark (namely B3, defined in Table 3.11) verifies the condition (3.154) and therefore the linearizations (L1) and (L2) coincide for the case at hand. Some relevant behaviours are shown in Figures 3.36 and 3.37, in which the label $L1/L2$ concisely refers to both the aforementioned linearizations while L_{orig} refers to the proposed one (3.130).

It is worth noticing that:

- for the considered test-case the considered discrete solutions turn out to be very similar to each other;

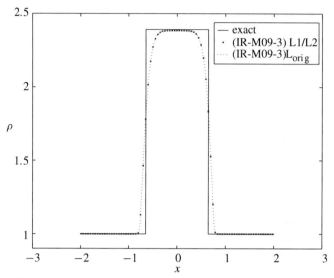

Figure 3.36. Comparison between the considered approximate linearizations: approximation of ρ (*i.e.* p) for the test-case IR-M09-3 in Table 3.12.

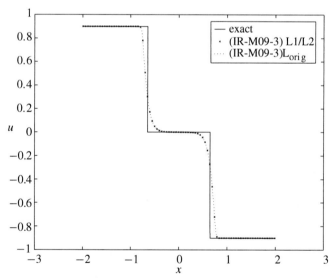

Figure 3.37. Comparison between the considered approximate linearizations: approximation of u for the test-case IR-M09-3 in Table 3.12.

- the stability of the linearized implicit time-advancing based on (L1) *i.e.* (L2) turns out to be comparable with that one based on the proposed linearization. Indeed, as far as the test-cases reported in Table 3.12 are concerned (hence, in particular for IR-M09-3), a blow-up occurs when advancing the simulations with a coefficient $\tilde{c}^{(CFL)} = O(10)$, namely for $1.5 \cdot 10^{-1} < \tau < 5.0 \cdot 10^{-1}$ (compare with the relevant point in Section 3.5.6).

Similar results have been obtained by exploiting a slightly different state law for which (L1) does not coincide with (L2). On the basis of the carried-out numerical experiments, the proposed linearization technique (3.130) seems to behave similarly to the aforementioned ones, as far as the accuracy and the efficiency are concerned. In particular, the stability restrictions imposed on the time-advancing by the presence of discontinuities within the flow field seem to affect all the considered linearized algorithms in a similar manner. However, only a preliminary investigation has been performed in this regard and further study is definitely recommended.

Chapter 4
1D Applications to cavitating flows

In Section 4.1 the considered state law is introduced, together with some details concerning its numerical implementation. In Section 4.2 some numerical results obtained in [91] are recalled and an illustrative numerical experiment, based on a RP whose IC leads to a cavitating flow, is presented.

4.1. State law of the working fluid

A state law of the type of that one defined in 1.5 is assumed for the working fluid. The density domain, in particular, is split into two adjacent subdomains: an upper one, where the working fluid behaves as a pure liquid, and a lower one, where cavitation phenomena are taken into account by a homogeneous flow cavitation model (see Sections 1.3 and 1.4).

The mathematical definition of both models, whose physical assumptions and implications are widely discussed in [26] and [27], is respectively given in Sections 4.1.1 and 4.1.2. Some issues regarding the numerical implementation of the cavitating mixture state law are then discussed in Section 4.1.3. Finally, in Section 4.1.4, the convexity of the chosen barotropic state law is discussed.

4.1.1. Pure liquid model

The working liquid is supposed to be at a constant temperature T_L. Let p_{sat} and ρ_{Lsat} be the saturation pressure and the liquid saturation density, respectively, at temperature T_L. Furthermore, let $\beta_L > 0$ denote the coefficient of isentropic compressibility of the liquid at temperature T_L [12]. The non-dimensional form of the considered liquid barotropic state law reads:

$$\bar{p} = \bar{p}_{\text{liq}}(\bar{\rho}) := 1 + \vartheta \ln(\bar{\rho}) \quad , \quad \bar{\rho} \in [1, \infty) \tag{4.1}$$

where:

$$\bar{\rho} := \frac{\rho}{\rho_{Lsat}} \quad , \quad \bar{p} := \frac{p}{p_{sat}} \quad , \quad \vartheta := (\beta_L \, p_{sat})^{-1} . \tag{4.2}$$

Note 4.1.1. Common liquids are nearly-incompressible: their non-dimensional compressibility coefficient ϑ, as computed by adopting physically-based values, is very high (*e.g.* $O\left(10^6\right)$ for water at 20°C). Consequently, the density is practically constant ($\bar{\rho} \approx 1$) for preventing unphysically high pressure values to be produced. In consideration of this point, in many computations involving real fluids under ordinary conditions, the logarithmic state law (4.1) is replaced with its linearization (see *e.g.* [96] and [78]), namely:

$$\bar{p} - 1 \approx \vartheta \, (\bar{\rho} - 1) \quad , \quad \bar{\rho} \in [1, \infty) . \tag{4.3}$$

The linearized state law may be preferable to the original one in view of its simplicity. It is worth noticing that the expression (4.3) represents a specific instance of the convex barotropic state law (2.66), obtained in particular for $\kappa = \vartheta$, $\varkappa = 1$ and $\gamma = 1 - \vartheta$.

4.1.2. Cavitation model

The chosen cavitation model provides the following differential relation between the non-dimensional density and pressure -introduced in (4.2)- within the mixture region:

$$\frac{\bar{p}}{\bar{\rho}} \frac{d\bar{\rho}}{d\bar{p}} = \bar{\rho} \left\{ (1 - \varepsilon) \, \vartheta^{-1} \, \bar{p} + \sigma_1 \, \varepsilon \, \bar{p}^{\sigma_2} \right\} + (1 - \bar{\rho}) \{ \sigma_3 \} , \quad \bar{\rho} < 1 \tag{4.4}$$

where:

- the non-dimensional parameters σ_1, σ_2 and σ_3 are defined as follows (p_{sat} being introduced in the previous section):

$$\sigma_1 := g^\star \left(\frac{p_c}{p_{sat}} \right)^\eta \quad , \quad \sigma_2 := -\eta \quad , \quad \sigma_3 := \frac{1}{\gamma_V}$$

in which p_c denotes the saturation pressure of the fluid at hand, γ_V represents the specific heat ratio (*i.e.* specific heat at constant pressure over specific heat at constant volume) of the relevant vapour and $g^\star$ and η are constants depending on the fluid under consideration [10]. It should be noticed, in particular, that σ_2 and σ_3 only depend on the chosen liquid while σ_1 is also affected by the liquid temperature T_L;

- the symbol ε denotes the following non-dimensional function of $\bar{\rho}$:

$$\varepsilon = \varepsilon_\zeta\,(\bar{\rho}) := \left\{ \left\{ ((1+\zeta)^3 - 1)\,\frac{1-\bar{\rho}}{\bar{\rho}} \right\}^{-3} + 1 \right\}^{-1/3}, \qquad \bar{\rho} < 1 \quad (4.5)$$

which describes the liquid volume fraction ($0 \leq \varepsilon \leq 1$) which is in thermal equilibrium with the cavities. The symbol $\zeta > 0$ in (4.5) denotes a free model parameter accounting for thermal cavitation effects and, possibly, for the concentration of the active cavitation nuclei [26, 27]. The function $\varepsilon_\zeta\,(\bar{\rho})$ is monotonically decreasing and, in particular, admits the following asymptotic behaviour:

$$\varepsilon_\zeta\,(\bar{\rho} \to 1) \to 0 \qquad (4.6)$$

which correctly models the fact that a negligible fraction of the liquid participates to the heat exchange at the interface with a vanishing cavity [26].

Note 4.1.2. It should be noticed that, once chosen the working liquid, the pure liquid state law (4.1) only depends on the chosen temperature T_L while the mixture model (4.4) also depends on the free parameter ζ.

The physical foundations of the considered cavitation model ensure, in particular, that the monotonicity requirement (1.4) is satisfied. Hence, once given T_L and ζ a value, the o.d.e. (4.4) can be numerically integrated with the following physically based initial condition:

$$\bar{p}\,(\bar{\rho} = 1) = 1\,. \qquad (4.7)$$

Moreover, due to some approximations that are introduced when deriving the cavitation model [26], the integration can only be extended down to a certain threshold $\bar{\rho}_{\min}$ such that:

$$\bar{\rho}_{\min} \gg \frac{\rho_{Vsat}}{\rho_{Lsat}} \qquad (4.8)$$

where ρ_{Vsat} is the vapour saturation density at temperature T_L. The condition (4.8) clearly prevents the model to be applied for describing liquid-vapour mixtures towards the pure vapour limit (hence, the chosen cavitation model is not suitable, as it is, for juxtaposition with a barotropic state law describing the pure vapour).

For consistency with the expression (1.1), the integral curve defined by (4.4) and (4.7) is formally denoted as follows:

$$\bar{p} = \bar{p}_{cav}\,(\bar{\rho}) \quad , \quad \bar{\rho} \in [\bar{\rho}_{\min}, 1) \qquad (4.9)$$

where the half-open density domain must be juxtaposed with that one of the pure liquid. The considered mixture state law (4.9), in particular, smoothly joins the liquid one (4.1) at the saturation point ($\bar{\rho} = 1$, $\bar{p} = 1$), up to the first derivative. Indeed, the continuity of p is trivially enforced by the initial condition (4.7). Moreover, by substituting (4.7) and (4.6) into (4.4), it follows that:

$$\frac{d\bar{\rho}}{d\bar{p}} (\bar{\rho} \to 1) \to \vartheta^{-1}$$

in agreement with the fact that, according to (4.1), $d\bar{p}/d\bar{\rho}\,(\bar{\rho} = 1) = \vartheta$. In other words, both p and a are continuous across the saturation point.

As an example, water at $T_L = 293.16$ K is considered, leading to the following values for the parameters in (4.4): $\vartheta \approx 8.55 \cdot 10^5$ (see [26] and [83]), $\sigma_1 \approx 1.33 \cdot 10^3$, $\sigma_2 \approx -0.73$ and $\sigma_3 \approx 0.78$ (see [10, 26] and [83]). Since $\rho_{Vsat}/\rho_{Lsat} = O\left(10^{-5}\right)$ for the case under consideration, a threshold $\bar{\rho}_{min} = O\left(10^{-4} \div 10^{-3}\right)$ is chosen in consideration of (4.8). Two barotropic curves obtained by choosing different values of the free parameter ζ are shown in Figure 4.1. The corresponding (dimensional) sound speed curves are reported in Figure 4.2. In this figure, the scale of the y-axis is deliberately cut for ease of readability. Indeed, both limits $a\,(\bar{\rho} \to 1) \to (\vartheta\,p_{sat}/\rho_{Lsat})^{1/2} \approx 1.41 \cdot 10^3$ m/s and $a\,(\bar{\rho} \to \bar{\rho}_{min}) \to O\left(10^2\right)$ m/s would in practice squash almost all the curve on the x-axis.

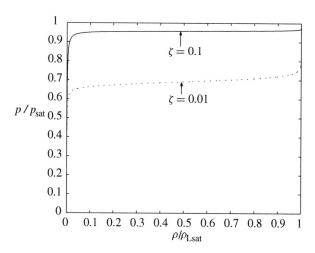

Figure 4.1. Typical trends of the considered mixture barotropic state law for water at $T_L = 293.16$ K.

The very sharp, step-like, transition of the sound speed occurring near the saturation point in Figure 4.2 is typical of the cavitation inception at low

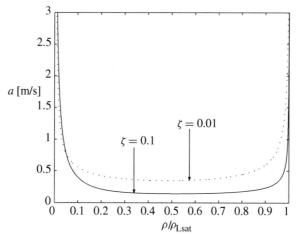

Figure 4.2. Trend of the mixture sound speed corresponding to the state laws shown in Figure 4.1. The y-axis is cut for ease of readability.

temperatures T_L ("cold cavitation"), of the type of the considered one. As already mentioned in Section 1.4, this abrupt transition is essentially related to the considered physical phenomenon, as modelled by a homogeneous cavitation model, and not to the specific model here adopted. In particular, it is also present when considering the well-known barotropic cavitation model originally proposed by Delannoy (see *e.g.* [22, 29] and [30]). Clearly, it is very challenging to incorporate state laws like those shown in Figure 4.1 (coupled with a suitable liquid model, *e.g.* (4.1) or (4.3)), into state-of-the-art numerical schemes.

4.1.3. Numerical implementation of the mixture state law

The cavitation model (4.4) does not explicitly provide the output of interest (*e.g.* p) in correspondence of the chosen input (*e.g.* ρ): to this purpose, an o.d.e. must be solved in advance. Obviously, when performing a simulation, it is not convenient from a computational standpoint to solve such an o.d.e. at each time-step and at each point of the computational grid in order to obtain the desired output. Thus, it seems convenient to numerically integrate the mixture cavitation model at the beginning of each simulation and then to store a table of the form:

$$(\rho_i , \ p_i , \ a_i) \quad , \quad i \in \{0, \ldots, n-1\} \tag{4.10}$$

with, say, $\rho_0 = \rho_{\text{Lsat}}$ and $\rho_{n-1} = \rho_{\text{min}} := \bar{\rho}_{\text{min}} \cdot \rho_{\text{Lsat}}$, to be accessed as required by the simulation algorithm. It is therefore necessary to define a fast table look-up strategy in order to efficiently incorporate the cavitating branch (4.9) of the barotropic model into a suitable numerical solver. To the purpose, it is possible to take advantage of the typical "S-like" shape of the mixture state law (see Figure 4.1), as explained below.

A density-based algorithm is assumed for the remainder of the present section. A typical access to the table (4.10) within such an algorithm is aimed at finding the pressure p and the sound speed a corresponding to a certain input value of the independent variable $\rho < \rho_{\text{Lsat}}$. It is possible to define a fast look-up strategy by firstly noticing that the distribution along the x-axis in Figure 4.1 of the density "nodes" ρ_i, as provided by an ordinary adaptive integration algorithm (*e.g.* a classical fourth-order Runge-Kutta scheme [79] with adaptive step-size control), typically exhibits clusters near the extremes $i = 0$ and $i = n-1$ due to the high value of the mixture sound speed, respectively in correspondence of $\rho = \rho_{\text{Lsat}}$ and $\rho = \rho_{\min}$. It is therefore possible to approximate the original density sequence ρ_i by a new one, say ρ'_j, obtained by juxtaposing two geometric sequences, $\rho_j^{(\text{right})}$ and $\rho_j^{(\text{left})}$, respectively starting from ρ_0 and ρ_{n-1} and joining each other at a certain node $\rho_{i_\star}$ such that $\rho_{i_\star} \approx 0.5 \cdot \rho_{\text{Lsat}}$. Let $\gamma_r > 1$ and $\gamma_l > 1$ denote the ratios of $\rho_j^{(\text{right})}$ and $\rho_j^{(\text{left})}$, respectively. Once defined the number of points in each sequence, say n_r and n_l respectively, the following representations are easily obtained:

$$\rho_j^{(\text{right})} := \rho_0 - \frac{\gamma_r^j - 1}{\gamma_r - 1}\, \delta_r, \quad j \in \{0, \ldots, (n_r - 1)\} \tag{4.11}$$

$$\rho_j^{(\text{left})} := \rho_{n-1} + \frac{\gamma_l^{(n_r+n_l-2)-j} - 1}{\gamma_l - 1}\, \delta_l, \quad j \in \{(n_r-1), \ldots, (n_r+n_l-2)\} \tag{4.12}$$

where:

$$\delta_r := \left(\rho_0 - \rho_{i_\star}\right) \frac{\gamma_r - 1}{\gamma_r^{(n_r-1)} - 1}$$

$$\delta_l := \left(\rho_{i_\star} - \rho_{n-1}\right) \frac{\gamma_l - 1}{\gamma_l^{(n_l-1)} - 1}$$

and the new density sequence finally reads:

$$\rho'_j := \begin{cases} \rho_0 & , \quad j = 0 \\ \rho_j^{(\text{right})} & , \quad j \in \{1, \ldots, (n_r - 2)\} \\ \rho_{i_\star} & , \quad j = (n_r - 1) \\ \rho_j^{(\text{left})} & , \quad j \in \{n_r, \ldots, (n_r + n_l - 3)\} \\ \rho_{n-1} & , \quad j = (n_r + n_l - 2). \end{cases} \tag{4.13}$$

The new density sequence has a noticeable advantage over the old one: it permits to analytically identify the nodal span to which a given value of the density ρ belongs by inverting (4.11) and (4.12) as follows (the cases $\rho = \rho_0$, $\rho = \rho_{i_\star}$ and $\rho = \rho_{n-1}$ are neglected because trivial):

$$
\rho \in
\begin{cases}
\left(\rho'_{\sigma(\rho)+1}, \rho'_{\sigma(\rho)} \right] & , \quad \rho_{i_\star} < \rho < \rho_0 \\[2ex]
\left[\rho'_{\tau(\rho)}, \rho'_{\tau(\rho)-1} \right) & , \quad \rho_{n-1} < \rho < \rho_{i_\star}
\end{cases}
\tag{4.14}
$$

with:

$$
\sigma(\rho) := \left\lfloor \frac{1}{\ln(\gamma_r)} \ln\left\{ 1 + (\rho_0 - \rho)\, \frac{\gamma_r - 1}{\delta_r} \right\} \right\rfloor
\tag{4.15}
$$

$$
\tau(\rho) := (n_r + n_l - 2) - \left\lfloor \frac{1}{\ln(\gamma_l)} \ln\left\{ 1 + (\rho - \rho_{n-1})\, \frac{\gamma_l - 1}{\delta_l} \right\} \right\rfloor
\tag{4.16}
$$

where, of course, the symbol $\lfloor \cdot \rfloor$ denotes the floor function.

Once defined the new density sequence ρ'_j, a new table can be built either by solving the o.d.e. (4.4) once more, now in correspondence of the sequence ρ'_j, or by interpolating the original table. The latter strategy is considered here and the following new table, in particular, is built:

$$
\left(\rho'_j, \ p'_j, \ a'_j \right) \quad , \quad j \in \{0, \ldots, (n_r + n_l - 2)\}
\tag{4.17}
$$

by linearly interpolating the original one (4.10) in correspondence of the new density sequence (4.13). Clearly, the original table can be discarded at this point, since it is never accessed by the considered algorithm. It may be worth noticing that, besides being attractive for its simplicity, a linear interpolation preserves the strict monotonicity of the p-ρ curve.

For suitable values of the relevant parameters, the new table very well approximates the original one, as shown for instance in Figure 4.3. It is therefore natural to define the following two-step access strategy based on the new table (4.17):

- given an input density ρ (the cases $\rho = \rho_0$, $\rho = \rho_{i_\star}$ and $\rho = \rho_{n-1}$ are not considered here because trivial), the corresponding span within the new table (4.17) is identified, by means of (4.14)-(4.16);
- the values of p and a corresponding to ρ are then defined by linear interpolation within the identified span. Of course, this procedure can be extended to an arbitrary number of dependent variables (*e.g.* the function Ψ, defined in (2.64), to be used for solving RPs associated with convex state laws, see Section 2.5.1).

Evidently, the aforementioned access strategy is more efficient than a crude look-up within the original table (4.10). A similar technique can be defined for pressure-based algorithms, as outlined in Appendix B.

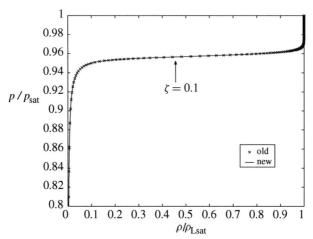

Figure 4.3. Comparison between the barotropic curves extracted from tables (4.10) ("old") and (4.17) ("new") for water at $T_L = 20°C$. Data: $n = 8127$, $i_* = 6586$, $\gamma_r = \gamma_l = 1.004$, $n_r = 6587$ and $n_l = 1541$. The y-axis is cut for ease of readability.

4.1.4. Convexity of the chosen state law

In consideration of the equality (2.54), the original convexity condition (2.53) is introduced as follows:

$$2a\ c(\rho) > 0$$

where $c(\rho)$ is defined by (2.55). Then, by substituting the expression of $c(\rho)$, the condition above is recast as follows, for later convenience:

$$2\frac{\varphi}{\rho} + \frac{d\varphi}{d\rho} > 0 \tag{4.18}$$

where:

$$\varphi := a^2(\rho).$$

The condition (4.18) is exploited below for assessing the convexity of the chosen barotropic state law.

For the pure liquid model (4.1)-(4.2) the following equality holds:

$$2\frac{\varphi}{\rho} + \frac{d\varphi}{d\rho} = \frac{1}{\beta_L\ \rho^2}$$

and therefore the convexity condition (4.18) is clearly satisfied ($\beta_L > 0$ by definition). This holds true also for the linearized liquid model (4.3), as already noticed in Note 4.1.1 (Section 4.1.1).

In view of the fact that $\varphi = a^2 = \mathrm{d}p/\mathrm{d}\rho$, the cavitation model (4.4) can be formally written in a more general way as follows:

$$\varphi = \frac{p}{\rho \; \omega(\rho, \, p(\rho))} \qquad (4.19)$$

where the function ω represents the right-hand side of (4.4). Then, by differentiating (4.19), the following equality is obtained:

$$2\frac{\varphi}{\rho} + \frac{\mathrm{d}\varphi}{\mathrm{d}\rho} = \frac{\varphi}{\rho} + \frac{\varphi^2}{p}\left\{1 - \rho\left(\frac{\partial\omega}{\partial\rho} + \varphi\,\frac{\partial\omega}{\partial p}\right)\right\}. \qquad (4.20)$$

The right-hand side of (4.20) can be exactly computed during the numerical integration of the o.d.e. (4.4) (of course, the partial derivatives of ω are known functions) and therefore it is possible to assess the convexity of the cavitating branch of specific state law as well.

Both the barotropic laws shown in Figure 4.1, for instance, turn out to be convex. Hence, in spite of the fact that the convexity marker $c(\rho)$ defined in (2.55) exhibits a jump at the junction between the liquid and the cavitating branch (which is due to the discontinuity of $\mathrm{d}a/\mathrm{d}\rho$ across the saturation point $(\rho_{\mathrm{Lsat}}, \, p_{\mathrm{sat}})$), the corresponding unified barotropic curve (*i.e.* (4.1) coupled with (4.9)) can be classified as convex on the whole [69]. However, the aforementioned discontinuity of $\mathrm{d}a/\mathrm{d}\rho$ across the saturation point is not a "pathology" affecting the chosen cavitation model. On the contrary, it reflects, within the limits of the homogeneous flow modelling, the characteristic behaviour of the state law at phase transition. Indeed, in general, "phase transitions in the fluid are a principal cause of non-convexity, since the sound speed in a mixed phase region is smaller than in the pure phase" [69].

It may be worth remarking that the approximations introduced when deriving homogeneous flow cavitation models may affect the convexity of the resulting state laws. Indeed, even small differences between two given homogeneous flow models can lead to substantially different wave solutions of the same system of governing equations. For instance, a non-convex barotropic state law is considered in [103], which is qualitatively similar to those shown in Figure 4.1. This law, which is smooth within the mixture region and which allows for smooth junctions with a pure liquid and a pure vapour barotropic models to be defined, is exploited in [103] to solve a RP by following [113]. Besides the classical rarefaction and shock waves presented in Section 2.4.2, so called "composite" waves appear as part of the solution, which are defined by juxtaposing up to three classical waves in an alternate fashion (*i.e.* shock-rarefaction, rarefaction-shock, shock-rarefaction-shock and rarefaction-shock-rarefaction.[1]) According to the author, the aforementioned sensitivity, besides highlighting the key role that modelling plays in this context, can encourage to also consider cavitation models which expressly take into account additional physical effects, *e.g.* non-homogeneous models (see Section 1.3) or models directly incorporating thermodynamic effects related to phase transition. This opinion seems to be somehow supported by the fact that difficulties arise in applying common entropy conditions (see Section 2.3.3) for selecting numerical solutions to classical p.d.e.s coupled with the state laws provided by classical homogeneous flow models (see *e.g.* [3] and [69]). Alternative approaches (*e.g.* the entropy-satisfying procedure based on the mixture thermodynamics which is proposed in [3]) should be carefully considered.

4.2. Numerical results

The Roe flux function, the preconditioning strategy and the linearized implicit time-advancing respectively presented in Sections 3.3, 3.4 and 3.5 have been originally introduced in [91]. A qualitative appraisal of the considered numerical ingredients is reported in the aforementioned document, based on the quasi-1D water flow within a convergent-divergent nozzle. In particular, the state law reported in Figure 4.1 which is associated with $\zeta = 0.1$ is considered in order to numerically simulate both non-cavitating and cavitating flows. The obtained results, simply recalled here for conciseness, show that:

- the semi-discrete scheme based on the proposed Roe flux function exhibits accuracy problems at the low Mach numbers typical of liquid flows. The considered preconditioning strategy effectively overcomes this problem (in particular, a local preconditioning strategy of the type of that one mentioned in Section 6.1.6 turns out to be effective also when cavitation occurs);
- the preconditioning technique restricts the stability of the considered explicit time-advancing algorithm (a 4-th order Runge-Kutta scheme). The proposed linearized implicit strategy counteracts this problem: it permits to efficiently advance in time the non-cavitating simulations. However, when cavitation takes place, a noticeable time-step restriction must be accepted; in particular, the allowable time-step turns out to be of the order of that one required by the explicit non-preconditioned scheme.

---

[1] No contact discontinuities are involved in the solution of the system at hand [103].

The relevant numerical experiments reported in Sections 3.3, 3.4 and 3.5 are in agreement with the aforementioned results. The time-step reduction which must be introduced when considering cavitating flows, in particular, is due to the occurrence of noticeable discontinuities -especially as far as the Mach number and the density are concerned- which are associated with the inception of cavitation (see Sections 1.4 and 4.1.2). However, as discussed in Section 3.5.7, this problem does not seem to be specifically introduced by the proposed linearization technique (3.130). Furthermore, it has been also observed by performing a rather extensive number of numerical simulations [7], based on the proposed linearization (3.130) and involving a different homogeneous flow cavitation model (namely the instance of the well-known barotropic cavitation model of Delannoy which is reported in [22]).

Clearly, once introduced a convex instance of the unified barotropic curve (4.1)-(4.9), it is possible to recall the material introduced in Section 2.5.3 in order to exactly solve 1D Riemann problems (RPs) based on the considered state law. These, in turn, provide exact benchmarks for validating 1D numerical methods dealing with cavitating flows and permit, in particular, to accurately investigate the behaviour of the considered numerical schemes at cavitation inception (thus addressing most of the difficulties related to the phase transition, as described by a homogeneous flow model). A systematic study of this type is postponed to a subsequent research stage; nevertheless, an illustrative test-case is considered in the sequel, showing some features that characterize the numerical discretization of the phase transition, as described by the unified barotropic model (4.1)-(4.9).

Benchmark

The considered benchmark is defined in Table 4.1. The mixture branch of the chosen state law is one of the two curves reported in Figure 4.1. The relevant non-dimensional dependent parameters for the expressions (4.1) and (4.4) are: $\vartheta \approx 8.55 \cdot 10^5$ (see [26] and [83]), $\sigma_1 \approx 1.33 \cdot 10^3$, $\sigma_2 \approx -0.73$ and $\sigma_3 \approx 0.78$ (see [10, 26] and [83]). At the chosen temperature T_L, the liquid saturation density is $\rho_{\mathrm{Lsat}} = 997.95$ and therefore the IC in Table 4.1 defines two liquid states[2] (passive scalars are neglected for the sake of simplicity). Moreover, the speeds u_L and u_R are chosen so as to obtain two rarefactions (symmetrical with respect to the original discontinuity $x = 0$) which lead to a cavitating star region characterized by $\rho_\star \approx 960.47 < \rho_{\mathrm{Lsat}}$ and $u_\star = 0$ (by symmetry). The sound speed in the liquid is $a_L = a_R \approx 1415.63$ while in the cavitating region it

---

[2] The SI units are tacitly understood, see Note 2.2.1 in Section 2.2.

falls down to approximately $a_{cav} \approx 0.37$; the resulting flow is entirely subsonic.[3] In consideration of the aforementioned variation of the sound speed, it is to be expected that the star region is hardly observable as part of the solution.

Table 4.1. Considered benchmark.

Benchmark	Liquid	T_L	ζ	ρ_L	u_L	ρ_R	u_R	t_{eval}
B4	water	293.16	0.1	998	−0.1	998	0.1	1

Discretization

The discrete scheme (3.132) is considered (more precisely, its basic-1D counterpart not involving the passive scalar ξ), associated with a uniform space discretization having measure $\mu = 1$ and a constant time-step τ. Transmissive BCs of the type of (3.28) are adopted, leading to equations similar to (3.143). The considered test-cases are reported in Table 4.2. The numerical approximation of ρ, p and u is shown in Figs. 4.4-4.8. It

Table 4.2. Considered test-cases.

Test-case	Benchmark	μ	(n_L, n_R)	τ
LdA1	B4	1	$(2, 2) \cdot 10^3$	10^{-4}
LdA2	B4	1	$(2, 2) \cdot 10^3$	10^{-3}
LdA3	B4	1	$(2, 2) \cdot 10^3$	10^{-2}

should be noticed that:

- the density undergoes a spike-like variation close to the cavitating region. It is practically impossible to distinguish the head as well as the tail of the density waves (see Note 8 in Section 3.2.2) in Figure 4.4, because the density variation close to rarefaction head is squashed by the considerable variation occurring towards the cavitating region. Furthermore, the width of the star region is not resolved by the adopted space discretization.

  When examining in Figure 4.5 a narrower sub-domain around $x = 0$ it is evident that, as expected, the accuracy of the numerical solution improves when adopting smaller time-steps. However, it is extremely difficult to accurately describe the cavity, which only occupies a very small region close to the minimum of the "exact" curve shown in Figure 4.5 (see the following point). Indeed, the characteristic size of

---

[3] Sonic conditions are deliberately avoided, see Note 3.3.12 in Section 3.3.1.

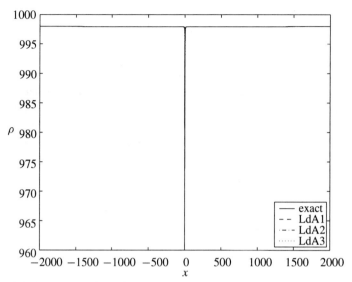

Figure 4.4. Approximation of ρ for the test-cases reported in Table 4.2.

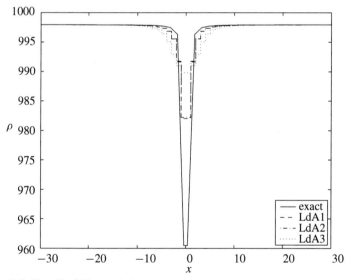

Figure 4.5. Detail of Figure 4.4.

the cavity is $O\left(a_{cav} \cdot t_{eval}\right) = O\left(10^{-1}\right)$, clearly finer than the adopted space discretization;[4]

---

[4] A uniform space discretization is adopted for consistency with the other 1D numerical experiments reported in the present document. A finer discretization is not considered in order not to introduce a computational overhead within the liquid region (which represents the vast majority of the computational domain).

- the pressure exhibits a remarkably different trend with respect to the density, as shown in Figure 4.6. Indeed:

    - the head of the rarefactions is clearly visible; that one of the left rarefaction, for example, is marked by P1 in the considered figure;
    - most of the pressure variation takes place, in practice, near the head of the rarefactions. For instance, as far as the left rarefaction is concerned, the pressure abruptly reaches the saturation value p_{sat}, marked by P2 in Figure 4.6, as well as the "right corner" of the relevant pressure curve in Figure 4.1 (very close to the right margin of the figure), marked by P3 in Figure 4.6. The transition between P2 and P3 is aligned with that one between P1 and P2 (*i.e.* no abrupt changes occur when entering the mixture region). Indeed, the rightmost portion of the relevant cavitating curve in Figure 4.1 is practically vertical near the saturation point (due to the very weak compressibility of the liquid) and therefore it behaves like a prolongation of the adopted liquid model;

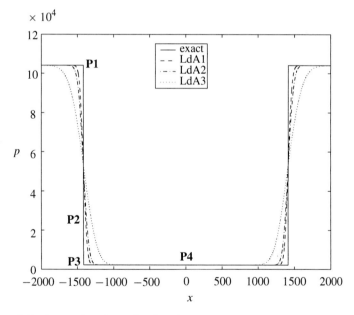

Figure 4.6. Approximation of p for the test-cases reported in Table 4.2. The labels P1-P4 are added for ease of discussion.

    - the cavity, whose width is $O\left(10^{-1}\right)$ (see above), is indicated in Figure 4.6 by P4. When moving from P3 to P4, the pressure weakly decreases (the variation is not resolved in the figure; the considered arc lies on the practically horizontal portion of the relevant pressure curve in Figure 4.1). A detail is shown in Figure 4.7;

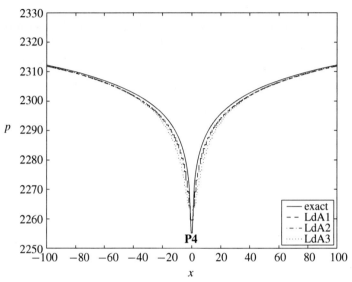

Figure 4.7. Detail of Figure 4.6.

- as shown in Figure 4.8, the approximation of u is reasonably good, even near the cavity.

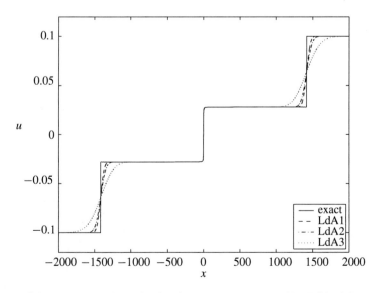

Figure 4.8. Approximation of u for the test-cases reported in Table 4.2.

In consideration of the previous points, it is clear that an accurate description of the ratrefaction's tail and, more in general, of the cavity is only possible at the cost of a very fine space discretization. Of course, several numerical investigations of the type of that one reported above can be performed by exploiting the chosen barotropic state law (the discretization of the sound speed, for instance, is considered in Figure 4.9). However, as stated above, such an investigation is postponed to a subsequent research stage.

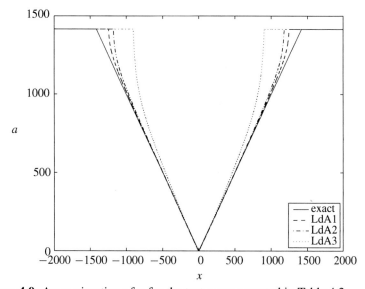

Figure 4.9. Approximation of a for the test-cases reported in Table 4.2.

Chapter 5
3D Numerical method

In the present chapter, a linearized implicit discrete scheme is proposed for solving the 3D governing equations introduced in Sections 2.2.1 and 2.2.2, based on some numerical ingredients introduced in Chapter 3. By adopting the architecture of the numerical frame mentioned in the introduction to the present document (namely the AERO code), the space and time discretizations are kept separate from each other.

As far as the space discretization is concerned, some basic issues regarding the considered unstructured grids are recalled in Section 5.1.1. Then, a generalization of the Roe numerical flux proposed in Section 3.3 is discussed in Section 5.1.2. Moreover, in Section 5.1.3 the preconditioning technique introduced in Section 3.4 is incorporated into the considered 3D Roe numerical flux. Finally, once specified the discretization of the convective fluxes, the relevant semi-discrete formulation is introduced in Section 5.1.4 and extended to rotating frames in Section 5.1.5.

As far as the time discretization is concerned, in Section 5.2.1 the linearization of the Roe numerical flux function proposed in Section 3.5 is generalized to the present 3D context. Furthermore, in Section 5.2.2 a linearized implicit discrete scheme is defined, based on the relevant material introduced in the preceding sections (numerical simulations exploiting this scheme are reported in Chapter 6).

5.1. Space discretization

In this section, the main issues regarding the adopted finite volume space discretization are discussed.

5.1.1. Finite volume approximation

The considered space discretization is based on the finite volume approach introduced in Section 3.1.1; the definition of the finite volume cells for the 3D case at hand is described below. At a preliminary stage, the considered 3D (bounded) computational domain $\mathcal{D} \in \mathbb{R}^3$ is approximated by means of a polyhedral domain $\mathcal{D}^{\text{pol}}$ which, in turn, is divided into N_t tetrahedra having vertices $\mathbf{P}_i$, with $i \in \mathcal{I} := \{1, \ldots, N_c\}$. Let T_h, with $h \in \mathcal{H} := \{1, \ldots, N_t\}$, denote the h-th tetrahedron; the following relations are (by construction) satisfied:

$$T_{h_1} \cap T_{h_2 \neq h_1} = \{0\} \quad , \quad \mathcal{D}^{\text{pol}} = \bigcup_{h \in \mathcal{H}} T_h \, .$$

The i-th finite volume cell C_i, associated with $\mathbf{P}_i$, is given by:

$$C_i = \bigcup_{h \in t(i)} C_i^{(h)}$$

where:

- $t(i) \subset \mathcal{H}$ is the set of indexes marking those tetrahedra which share $\mathbf{P}_i$ as a vertex;
- $C_r^{(h)}$ represents the subset of T_h which is defined by further dividing T_h into 24 sub-tetrahedra by means of its median planes[1] and subsequently considering those 6 sub-tetrahedra which share $\mathbf{P}_r$ as a vertex.

Clearly, there is a finite volume cell for each vertex.[2] Moreover, the resulting finite volume discretization clearly verifies the following relations:

$$C_{i_1} \cap C_{i_2 \neq i_1} = \{0\} \quad , \quad \mathcal{D}^{\text{pol}} = \bigcup_{i \in \mathcal{I}} C_i$$

and it is sometimes referred to as a "dual mesh" (see *e.g.* [39]), by virtue of the specific procedure which is adopted in order to build the cells by starting from the tetrahedra.

An example of the construction of the finite volume cells is shown in Figure 5.1, for the 2D counterpart of the aforementioned 3D case. In this figure the tetrahedra are replaced with triangles in the $x_1 - x_2$ plane (whose vertices and edges are respectively marked by circles and dashed lines) and the median planes reduce to the ordinary medians (marked by dotted lines). Each triangle is then divided into 6 sub-triangles by the medians and 2 of them are associated with each vertex. The boundary of the cell C_i associated with $\mathbf{P}_i$ is identified by a solid line and the portion of this boundary representing, in particular, the interface between C_i and C_h is highlighted by a thicker line.

---

[1] Each median plane is associated with an edge. The median plane relative to a certain edge $\bar{e}$ contains $\bar{e}$ as well as the middle point of the (unique) edge $\tilde{e}$ which is not directly connected to $\bar{e}$.

[2] The considered finite volume discretization can be regarded to as a "cell vertex" one [39], even if it is not necessary -to the purposes of the present study- to associate the quantities defined on C_i with a specific point belonging to C_i (in particular with $\mathbf{P}_i$).

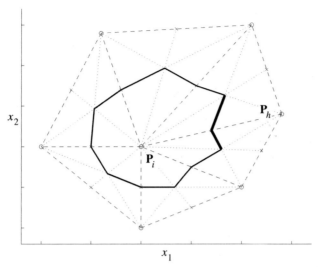

Figure 5.1. Example of the construction of a 2D finite volume cell by a dual mesh approach based on the medians.

Let μ_i represent the measure of C_i. On C_i the exact solution $\mathbf{q}(\mathbf{x}, t)$, where $\mathbf{x} \in \mathbb{R}^3$ denotes the position vector and $\mathbf{q}$ represents the conservative state vector defined in (2.4), is approximated by a semi-discrete function $\mathbf{q}_i(t)$ which is considered as an approximation of the mean value of $\mathbf{q}(\mathbf{x}, t)$ over C_i (in analogy with (3.1)):

$$\mathbf{q}_i(t) \approx \frac{1}{\mu_i} \int_{C_i} \mathbf{q}(\mathbf{x}, t) \, \mathrm{dV}. \tag{5.1}$$

The differential system defining $\mathbf{q}_i$ is obtained by discretizing the integral balance (2.3) over the control volume C_i. To the purpose, by virtue of (5.1), the time-derivative in (2.3) is naturally approximated as follows:

$$\partial_t \int_{C_i} \mathbf{q}(\mathbf{x}, t) \, \mathrm{dV} \approx \mu_i \frac{\mathrm{d}}{\mathrm{d}t} \mathbf{q}_i \tag{5.2}$$

while the term involving the flux is discretized as described in Section 5.1.4.

5.1.2. A 3D Roe numerical flux for generic barotropic state laws

In the spirit of (3.4), let:

$$\boldsymbol{\phi} \left(\mathbf{q}_i, \mathbf{q}_j, \hat{\boldsymbol{v}}_{ij} \right) \tag{5.3}$$

denote a 3D numerical flux from $\mathbf{q}_i$ to $\mathbf{q}_j$, along the direction $\hat{\boldsymbol{v}}_{ij}$. An instance of the aforementioned flux function is defined in the present section, based on the proposed augmented-1D Roe numerical flux (3.64)-(3.68).

Frame change and rotational invariance

Let $\mathbf{R} \in \mathbb{R}^{3 \times 3}$ denote an orthogonal matrix ($\mathbf{R}^{-1} = \mathbf{R}^T$) associated with a rotation of the chosen Cartesian frame (see Section 2.2.1). By introducing a matrix $\bar{\mathbf{R}} \in \mathbb{R}^{4 \times 4}$ defined as follows:

$$\bar{\mathbf{R}} := \begin{pmatrix} 1 & \mathbf{0}^T \\ \mathbf{0} & \mathbf{R} \end{pmatrix}$$

it is straightforward to compactly apply the aforementioned rotation to the considered state vector $\mathbf{q} \in \mathbb{R}^4$ (whose first component is a scalar, obviously invariant with respect to a frame change) as follows:

$$\mathbf{q} \longrightarrow \mathbf{q}' := \bar{\mathbf{R}} \cdot \mathbf{q}.$$

In order to correctly discretize the considered balance (2.3), which is properly formulated as a tensorial relation, the flux function (5.3) must satisfy the following property (rotational invariance, see *e.g.* [39] and [98]):

$$\boldsymbol{\phi}\left(\mathbf{q}_i, \mathbf{q}_j, \hat{\boldsymbol{v}}_{ij}\right) = \bar{\mathbf{R}}^{-1} \cdot \boldsymbol{\phi}\left(\mathbf{q}'_i, \mathbf{q}'_j, \hat{\boldsymbol{v}}'_{ij}\right) \tag{5.4}$$

where, of course, $\hat{\boldsymbol{v}}'_{ij}$ corresponds in the rotated frame to $\hat{\boldsymbol{v}}_{ij}$:

$$\hat{\boldsymbol{v}}'_{ij} := \mathbf{R} \cdot \hat{\boldsymbol{v}}_{ij}.$$

Sweep approximation

A frame rotation $\mathbf{R}$ is considered; without any loss of generality, the rotated direction $\hat{\boldsymbol{v}}'_{ij}$ is supposed to coincide with the versor $\hat{\mathbf{e}}'^{(k)}$ associated with the k-th direction x'_k ($k \in \{1, 2, 3\}$) of the rotated frame:

$$\hat{\boldsymbol{v}}'_{ij} = \hat{\mathbf{e}}'^{(k)}. \tag{5.5}$$

Moreover, a basic-1D flow is assumed to take place along x'_k (this assumption plays a fundamental role in the subsequent derivation). Thus, by recalling the k-th sweep of the relevant 3D governing equations (see Section 2.2.5) written in the rotated frame, it is possible to define an instance of the flux function $\boldsymbol{\phi}(\mathbf{q}'_i, \mathbf{q}'_j, \hat{\boldsymbol{v}}'_{ij})$ appearing in (5.4). More precisely, due to the formal identity between the augmented-1D equations and the 1D sweeps of the 3D equations (see Section 2.2.5), the considered instance of $\boldsymbol{\phi}(\mathbf{q}'_i, \mathbf{q}'_j, \hat{\boldsymbol{v}}'_{ij})$ can be defined by introducing a Roe numerical flux $\boldsymbol{\phi}'^{ROE}_{ij}$ which generalizes the proposed one (3.64)-(3.68), as described below.

The centred component of $\phi_{ij}^{\prime\text{ROE}}$ is firstly considered. Let $\mathbf{f}^{(\hat{v}'_{ij})}(\mathbf{q}')$ denote the augmented-1D analytical flux along x'_k. In consideration of (5.5), it is straightforward to derive from (2.5) the following representation:

$$\mathbf{f}^{(\hat{v}'_{ij})}\left(\mathbf{q}'\right) = \left(\mathbf{u}'^T \cdot \hat{v}'_{ij}\right) \mathbf{q}' + p \begin{pmatrix} 0 \\ \hat{v}'_{ij} \end{pmatrix} \tag{5.6}$$

where, of course:

$$\mathbf{u}' := \mathbf{R} \cdot \mathbf{u}.$$

The sought centred component, which generalizes the augmented-1D one (3.66), can then be defined as follows:

$$\phi_{c,ij}^{\prime\text{ROE}} := \frac{1}{2} \left(\mathbf{f}^{(\hat{v}'_{ij})}\left(\mathbf{q}'_i\right) + \mathbf{f}^{(\hat{v}'_{ij})}\left(\mathbf{q}'_j\right) \right).$$

The upwind component of $\phi_{ij}^{\prime\text{ROE}}$ is considered in the sequel. According to the sweep approximation, the velocity components associated with the versors $\hat{\mathbf{e}}'^{(h)}$, $h \neq k$, of the rotated frame are treated as passive scalars. Consequently, the Roe averages to be introduced in a Roe matrix for the k-th sweep under consideration are a_{ij}, defined in (3.62), and $\mathbf{u}'_{ij}$, with $\mathbf{u}'_{ij}$ defined as follows:

$$\mathbf{u}'_{ij} := \mathbf{R} \cdot \mathbf{u}_{ij} \tag{5.7}$$

where:

$$\mathbf{u}_{ij} := \frac{\sqrt{\rho_i}\,\mathbf{u}_i + \sqrt{\rho_j}\,\mathbf{u}_j}{\sqrt{\rho_i} + \sqrt{\rho_j}}. \tag{5.8}$$

The above definition, in particular, extends the Roe averages (3.61) to the present context. Let $\tilde{\mathbf{J}}'_{ij}$ denote the sought Roe matrix, which clearly generalizes the matrix $s_{ij}\,\tilde{\mathbf{J}}_{ij}^{(A)}$ appearing in (3.68). By a straightforward extension of (3.58), it is possible to define $\tilde{\mathbf{J}}'_{ij}$ by evaluating the Jacobian associated with the direction x'_k -which, in turn, can be derived from (2.11)- in correspondence of the aforementioned Roe averages, namely:

$$\tilde{\mathbf{J}}'_{ij} := \begin{pmatrix} 0 & \hat{v}'^T_{ij} \\ a^2_{ij}\,\hat{v}'_{ij} - \sigma_{ij}\,\mathbf{u}'_{ij} & \mathbf{u}'_{ij} \cdot \hat{v}'^T_{ij} + \sigma_{ij}\,\mathbf{I} \end{pmatrix} \tag{5.9}$$

where:

$$\sigma_{ij} := \mathbf{u}'^T_{ij} \cdot \hat{v}'_{ij} = \mathbf{u}^T_{ij} \cdot \hat{v}_{ij}. \tag{5.10}$$

Then, once recalled the definition of Δ^{ij} given in (3.63), the sought upwind component generalizing (3.67)-(3.68) can be defined as follows:

$$\phi_{u,ij}^{'ROE} \ := \ \mathbf{D}_{ij}' \cdot \Delta^{ij} \mathbf{q}' \tag{5.11}$$

$$\mathbf{D}_{ij}' \ := \ -\frac{1}{2} \left| \tilde{\mathbf{J}}_{ij}' \right|$$

and the resulting Roe flux function:

$$\phi_{ij}^{'ROE} \ := \ \phi_{c,ij}^{'ROE} + \phi_{u,ij}^{'ROE} \tag{5.12}$$

can be considered as an instance of the flux function $\phi(\mathbf{q}_i',\mathbf{q}_j',\hat{\mathbf{v}}_{ij}')$, namely:

$$\phi\left(\mathbf{q}_i', \mathbf{q}_j', \hat{\mathbf{v}}_{ij}'\right) = \phi_{ij}^{'ROE} . \tag{5.13}$$

3D Roe numerical flux

In view of (5.13), the representation of the considered instance of $\phi\left(\mathbf{q}_i, \mathbf{q}_j, \hat{\mathbf{v}}_{ij}\right)$ can be derived from the definitions introduced in the previous paragraph, by a trivial change of notation. Nevertheless, such a representation is reported below for ease of presentation.

Let the considered 3D Roe numerical flux function ϕ_{ij}^{ROE} be defined as follows:

$$\phi_{ij}^{ROE} \ := \ \phi_{c,ij}^{ROE} + \phi_{u,ij}^{ROE} \tag{5.14}$$

$$\phi_{c,ij}^{ROE} \ := \ \frac{1}{2}\left(\mathbf{f}^{(\hat{v}_{ij})}\left(\mathbf{q}_i\right) + \mathbf{f}^{(\hat{v}_{ij})}\left(\mathbf{q}_j\right)\right) \tag{5.15}$$

$$\phi_{u,ij}^{ROE} \ := \ \mathbf{D}_{ij} \cdot \Delta^{ij} \mathbf{q}$$

$$\mathbf{D}_{ij} \ := \ -\frac{1}{2}\left|\tilde{\mathbf{J}}_{ij}\right| \tag{5.16}$$

where the function $\mathbf{f}^{(\hat{v}_{ij})}\left(\mathbf{q}\right)$ in (5.15) is straightforwardly derived from (5.6) as follows:

$$\mathbf{f}^{(\hat{v}_{ij})}\left(\mathbf{q}\right) = \left(\mathbf{u}^T \cdot \hat{\mathbf{v}}_{ij}\right)\mathbf{q} + p\begin{pmatrix} 0 \\ \hat{\mathbf{v}}_{ij} \end{pmatrix} \tag{5.17}$$

and the Roe matrix $\tilde{\mathbf{J}}_{ij}$ in (5.16) is trivially derived from (5.9) as follows:

$$\tilde{\mathbf{J}}_{ij} := \begin{pmatrix} 0 & \hat{\mathbf{v}}_{ij}^T \\ a_{ij}^2\,\hat{\mathbf{v}}_{ij} - \sigma_{ij}\,\mathbf{u}_{ij} & \mathbf{u}_{ij} \cdot \hat{\mathbf{v}}_{ij}^T + \sigma_{ij}\,\mathbf{I} \end{pmatrix} \tag{5.18}$$

with σ_{ij} introduced in (5.10). From (5.13) it follows that the considered instance of the 3D flux function (5.3) reads:

$$\phi\left(\mathbf{q}_i, \mathbf{q}_j, \hat{\mathbf{v}}_{ij}\right) = \phi_{ij}^{\text{ROE}} \qquad (5.19)$$

with ϕ_{ij}^{ROE} defined in (5.14)-(5.18).

Note 5.1.1. The numerical flux (5.19) evidently satisfies the following relations:

$$\phi\left(\mathbf{q}_j, \mathbf{q}_i, \hat{\mathbf{v}}_{ji} = -\hat{\mathbf{v}}_{ij}\right) = -\phi\left(\mathbf{q}_i, \mathbf{q}_j, \hat{\mathbf{v}}_{ij}\right) \qquad (5.20)$$

$$\phi\left(\mathbf{q}_i, \mathbf{q}_j = \mathbf{q}_i, \hat{\mathbf{v}}_{ij}\right) = \mathbf{f}^{(\hat{v}_{ij})}\left(\mathbf{q}_i\right) \qquad (5.21)$$

with $\mathbf{f}^{(\hat{v}_{ij})}\left(\mathbf{q}\right)$ given by (5.17). The relation (5.20) clearly extends the conservation property (3.7) while the relation (5.21) provides a generalization of the consistency property (3.8).

Moreover, the numerical flux (5.19) also satisfies the rotational invariance condition (5.4). Clearly, in order to verify the previous assertion it suffices to show that the following relation:

$$\phi_{ij}^{\prime\text{ROE}} = \bar{\mathbf{R}} \cdot \phi_{ij}^{\text{ROE}} \qquad (5.22)$$

holds true. To the purpose, the centred components are firstly considered. Once noticed that the right-hand side of the expression (5.6) can be recast as follows (of course, $\mathbf{u}^{\prime T} \cdot \hat{\mathbf{v}}_{ij}^{\prime} = \mathbf{u}^T \cdot \hat{\mathbf{v}}_{ij}$):

$$\left(\mathbf{u}^T \cdot \hat{\mathbf{v}}_{ij}\right) \bar{\mathbf{R}} \cdot \mathbf{q} + p\,\bar{\mathbf{R}} \cdot \begin{pmatrix} 0 \\ \hat{\mathbf{v}}_{ij} \end{pmatrix} = \bar{\mathbf{R}} \cdot \mathbf{f}^{(\hat{v}_{ij})}\left(\mathbf{q}\right)$$

with $\mathbf{f}^{(\hat{v}_{ij})}\left(\mathbf{q}\right)$ given by (5.17), it is evident that:

$$\phi_{c,ij}^{\prime\text{ROE}} = \bar{\mathbf{R}} \cdot \phi_{c,ij}^{\text{ROE}}. \qquad (5.23)$$

As far as the upwind components are concerned, the following relation (straightforwardly derived from the relevant definitions) can be introduced:

$$\tilde{\mathbf{J}}_{ij}^{\prime} = \bar{\mathbf{R}} \cdot \tilde{\mathbf{J}}_{ij} \cdot \bar{\mathbf{R}}^{-1}. \qquad (5.24)$$

In consideration of the fact that the eigenvalues are invariant with respect to a frame change and by recalling the definition of the operator $|\cdot|$ given in (2.1), it follows from (5.24) that:

$$|\tilde{\mathbf{J}}_{ij}^{\prime}| = \bar{\mathbf{R}} \cdot |\tilde{\mathbf{J}}_{ij}| \cdot \bar{\mathbf{R}}^{-1} \quad \Rightarrow \quad \mathbf{D}_{ij}^{\prime} = \bar{\mathbf{R}} \cdot \mathbf{D}_{ij} \cdot \bar{\mathbf{R}}^{-1}.$$

Hence, the right-hand side of (5.11) can be recast as follows (of course, $\Delta^{ij}\mathbf{q}' = \bar{\mathbf{R}} \cdot \Delta^{ij}\mathbf{q}$):

$$\bar{\mathbf{R}} \cdot \mathbf{D}_{ij} \cdot \bar{\mathbf{R}}^{-1} \cdot \bar{\mathbf{R}} \cdot \Delta^{ij}\mathbf{q} = \bar{\mathbf{R}} \cdot \mathbf{D}_{ij} \cdot \Delta^{ij}\mathbf{q}$$

and therefore:

$$\boldsymbol{\phi}_{u,ij}^{\prime\mathrm{ROE}} = \bar{\mathbf{R}} \cdot \boldsymbol{\phi}_{u,ij}^{\mathrm{ROE}}. \tag{5.25}$$

As a result, the equality (5.22) immediately follows from (5.23) and (5.25), in view of the definitions (5.12) and (5.14).

5.1.3. Incorporation of the preconditioning strategy

It is possible to extend the preconditioning strategy introduced in Section 3.4.3 so as to be incorporated into the proposed 3D Roe numerical flux (5.14)-(5.18). To the purpose, the preconditioner (3.102) is firstly recalled. Consistently with the sweep approximation introduced in the previous section, the representation of the preconditioner $\mathbf{P}'_{ij}$ -to be incorporated into the Roe flux $\boldsymbol{\phi}_{ij}^{\prime\mathrm{ROE}}$ defined in (5.12)- can be derived from the expression (3.102) by replacing $\left(u_{ij}, \xi_{ij}, \eta_{ij}\right)^T$ with the Roe averages (5.7), namely:

$$\mathbf{P}'_{ij} := \mathbf{I} + \left(\beta^2 - 1\right) \begin{pmatrix} 1 & \mathbf{0}^T \\ \mathbf{u}'_{ij} & \mathbf{O} \end{pmatrix}. \tag{5.26}$$

The matrix (5.26) satisfies the following relation:

$$\mathbf{P}'_{ij} = \bar{\mathbf{R}} \cdot \mathbf{P}_{ij} \cdot \bar{\mathbf{R}}^{-1}$$

where, of course:

$$\mathbf{P}_{ij} := \mathbf{I} + \left(\beta^2 - 1\right) \begin{pmatrix} 1 & \mathbf{0}^T \\ \mathbf{u}_{ij} & \mathbf{O} \end{pmatrix} \tag{5.27}$$

with the Roe averages $\mathbf{u}_{ij}$ defined in (5.8). The preconditioner (5.27) must be associated with the Roe matrix $\tilde{\mathbf{J}}_{ij}$ defined in (5.18) and the resulting 3D preconditioned Roe numerical flux finally reads:

$$\boldsymbol{\phi}_{ij}^{\mathrm{ROE,p}} := \boldsymbol{\phi}_{c,ij}^{\mathrm{ROE}} + \boldsymbol{\phi}_{u,ij}^{\mathrm{ROE,p}} \tag{5.28}$$

where the centred component $\boldsymbol{\phi}_{c,ij}^{\mathrm{ROE}}$ is given by (5.15) while the upwind one reads:

$$\boldsymbol{\phi}_{u,ij}^{\mathrm{ROE,p}} := \mathbf{D}_{ij}^p \cdot \Delta^{ij}\mathbf{q}$$

$$\mathbf{D}_{ij}^p := -\frac{1}{2} \left(\mathbf{P}_{ij}\right)^{-1} \cdot \left|\mathbf{P}_{ij} \cdot \tilde{\mathbf{J}}_{ij}\right|. \tag{5.29}$$

It may be worth remarking that, as for the starting augmented-1D case, the matrix $\mathbf{P}_{ij} \cdot \tilde{\mathbf{J}}_{ij}$ appearing in (5.29) is diagonalizable with real eigenvalues and therefore the operator $| \cdot |$ -defined in (2.1)- can be rightfully applied. The Roe numerical flux (5.28)-(5.29) can be considered as an instance of the 3D flux function (5.3); for later convenience, it is marked as follows:

$$\boldsymbol{\phi}^p \left(\mathbf{q}_i, \mathbf{q}_j, \hat{\mathbf{v}}_{ij} \right) := \boldsymbol{\phi}_{ij}^{\text{ROE},p} . \tag{5.30}$$

5.1.4. Discretization of the fluxes and semi-discrete formulation

The discretization of the surface integral appearing in the balance (2.3) is considered in the present section. More precisely, the convective flux across the boundary ∂C_i of the generic cell C_i is considered.

In general, the relevant integrand can be recast as follows:

$$\sum_{k=1}^{3} \hat{n}_k \, \mathbf{f}^{(k)} = \mathbf{f}^{(\hat{\mathbf{n}})}$$

where the definition of $\mathbf{f}^{(\hat{\mathbf{n}})} \, (\cdot)$ is trivially derived from (5.17), namely:

$$\mathbf{f}^{(\hat{\mathbf{n}})} \, (\mathbf{q}) := \left(\mathbf{u}^T \cdot \hat{\mathbf{n}} \right) \mathbf{q} + p \begin{pmatrix} 0 \\ \hat{\mathbf{n}} \end{pmatrix} . \tag{5.31}$$

As far as the integration domain is concerned, it can be split into several parts, as described below. The cell C_i is adjacent to a certain number $s(i)$ of other cells C_j, clearly equal to the number of vertices $\mathbf{P}_j$ which are connected to $\mathbf{P}_i$ by an edge of the underlying tetrahedral lattice (see Section 5.1.1). Consequently, the boundary ∂C_i of C_i can be decomposed as follows:[3]

$$\partial C_i = \left(\bigcup_{j \in s(i)} \partial C_i \cap \partial C_j \right) \cup \left(\partial C_i \cap \partial \mathcal{D}^{\text{pol}} \right) \tag{5.32}$$

where $\partial \mathcal{D}^{\text{pol}}$ denotes the boundary of the flow domain $\mathcal{D}^{\text{pol}}$. This boundary, in turn, is assumed to be split as follows:

$$\partial \mathcal{D}^{\text{pol}} = \partial \mathcal{D}^I \cup \partial \mathcal{D}^O \cup \partial \mathcal{D}^B \cup \partial \mathcal{D}^C \tag{5.33}$$

where $\partial \mathcal{D}^I$ and $\partial \mathcal{D}^O$ respectively denote the inflow and the outflow surfaces, $\partial \mathcal{D}^B$ represents the wall of a rigid body immersed within the flow

---

[3] A detailed characterization of the considered boundary can be found in [33].

(if any) and $\partial \mathcal{D}^C$ indicates a rigid wall encasing the flow. Consequently, once introduced the following definitions:

$$\mathcal{S}_{ij} := \partial C_i \cap \partial C_j \quad , \quad \boldsymbol{\varphi}_{ij} := \int_{\mathcal{S}_{ij}} \mathbf{f}^{(\hat{\mathbf{n}})} \, dS$$

$$\mathcal{S}_{iX} := \partial C_i \cap \partial \mathcal{D}^X \quad , \quad \boldsymbol{\varphi}_{iX} := \int_{\mathcal{S}_{iX}} \mathbf{f}^{(\hat{\mathbf{n}})} \, dS \quad , \quad X \in \{I, O, B, C\}$$

the convective flux across the boundary ∂C_i in the balance (2.3) can be recast as follows:

$$\int_{\partial C_i} \mathbf{f}^{(\hat{\mathbf{n}})} \, dS = \sum_{j \in s(i)} \boldsymbol{\varphi}_{ij} + \boldsymbol{\varphi}_{iI} + \boldsymbol{\varphi}_{iO} + \boldsymbol{\varphi}_{iB} + \boldsymbol{\varphi}_{iC} . \tag{5.34}$$

The discretization of each flux appearing on the right-hand side of (5.34) is discussed below.

In order to define a numerical approximation $\tilde{\boldsymbol{\varphi}}_{ij}$ of $\boldsymbol{\varphi}_{ij}$, the following average direction $\hat{\boldsymbol{v}}_{ij}$ associated with $\mathcal{S}_{ij}$ is introduced:

$$\boldsymbol{v}_{ij} := \int_{\mathcal{S}_{ij}} \hat{\mathbf{n}} \, dS \quad , \quad \hat{\boldsymbol{v}}_{ij} := \frac{\boldsymbol{v}_{ij}}{\|\boldsymbol{v}_{ij}\|} \tag{5.35}$$

and $\boldsymbol{\varphi}_{ij}$ is firstly approximated by a 3D numerical flux of the type of (5.3) crossing an "equivalent" planar surface having measure $\|\boldsymbol{v}_{ij}\|$ and normal $\hat{\boldsymbol{v}}_{ij}$, namely:

$$\boldsymbol{\varphi}_{ij} \approx \|\boldsymbol{v}_{ij}\| \, \boldsymbol{\phi} \left(\mathbf{q}_i, \mathbf{q}_j, \hat{\boldsymbol{v}}_{ij} \right) .$$

In consideration of the fact that the convective flux along $\hat{\boldsymbol{v}}_{ij}$, as obtained by substituting $\hat{\mathbf{n}} = \hat{\boldsymbol{v}}_{ij}$ into (5.31), coincides with the expression (5.17), it is possible to choose the proposed 3D Roe flux $\boldsymbol{\phi}_{ij}^{\mathrm{ROE}}$ defined in (5.14)-(5.18) for approximating $\boldsymbol{\varphi}_{ij}$. More in general, the 3D preconditioned Roe flux $\boldsymbol{\phi}_{ij}^{\mathrm{ROE,p}}$ given in (5.28)-(5.29) can be considered and therefore, in view of (5.30), the following approximation is defined:

$$\boldsymbol{\varphi}_{ij} \approx \tilde{\boldsymbol{\varphi}}_{ij} := \|\boldsymbol{v}_{ij}\| \, \boldsymbol{\phi}^p \left(\mathbf{q}_i, \mathbf{q}_j, \hat{\boldsymbol{v}}_{ij} \right) . \tag{5.36}$$

An approximation of the type of (5.36) is also adopted for the fluxes $\boldsymbol{\varphi}_{iI}$ and $\boldsymbol{\varphi}_{iO}$. More precisely, once introduced a fictitious inflow state vector $\mathbf{q}_i^I$, the following relation is introduced:

$$\boldsymbol{\varphi}_{iI} \approx \tilde{\boldsymbol{\varphi}}_{iI} := \|\boldsymbol{v}_{iI}\| \, \boldsymbol{\phi}^p \left(\mathbf{q}_i, \mathbf{q}_i^I, \hat{\boldsymbol{v}}_{iI} \right) \tag{5.37}$$

where $\hat{\boldsymbol{v}}_{iI}$ is defined in the spirit of (5.35). Similarly, the chosen approximation of the outflow flux reads:

$$\boldsymbol{\varphi}_{iO} \approx \tilde{\boldsymbol{\varphi}}_{iO} := \|\boldsymbol{v}_{iO}\| \, \boldsymbol{\phi}^p \left(\mathbf{q}_i, \mathbf{q}_i^O, \hat{\boldsymbol{v}}_{iO} \right) \tag{5.38}$$

where $\mathbf{q}_i^O$ represents a fictitious outflow state vector and $\hat{\mathbf{v}}_{iO}$ is defined in the spirit of (5.35). It should be noticed that the approximations (5.37) and (5.38), besides being consistent with the discretization of the inner fluxes (5.36), take into account the wave structure of the flow entering/exiting the computational domain by means of the upwinding component of the considered Roe numerical flux function.

At the walls $\partial \mathcal{D}^B$ and $\partial \mathcal{D}^C$ the classical slip condition [88]:

$$\mathbf{u}^T \cdot \hat{\mathbf{n}} = \mathbf{0} \tag{5.39}$$

is imposed, consistently with the adopted inviscid approximation (see Section 2.2). The condition (5.39) can be introduced into (5.31), thus leading to the following approximations:

$$\varphi_{iB} \approx \tilde{\varphi}_{iB} := \|\mathbf{v}_{iB}\| \begin{pmatrix} 0 \\ p_i \, \hat{\mathbf{v}}_{iB} \end{pmatrix} \tag{5.40}$$

$$\varphi_{iC} \approx \tilde{\varphi}_{iC} := \|\mathbf{v}_{iC}\| \begin{pmatrix} 0 \\ p_i \, \hat{\mathbf{v}}_{iC} \end{pmatrix} \tag{5.41}$$

where $\hat{\mathbf{v}}_{iB}$ and $\hat{\mathbf{v}}_{iC}$ are clearly defined in the spirit of (5.35).

In consideration of the material introduced in the present section, the convective flux (5.34) is discretized as follows:

$$\int_{\partial C_i} \mathbf{f}^{(\hat{n})} \, dS \approx \tilde{\varphi}_i := \sum_{j \in s(i)} \tilde{\varphi}_{ij} + \tilde{\varphi}_{iI} + \tilde{\varphi}_{iO} + \tilde{\varphi}_{iB} + \tilde{\varphi}_{iC} \tag{5.42}$$

with $\tilde{\varphi}_{ij}$, $\tilde{\varphi}_{iI}$, $\tilde{\varphi}_{iO}$, $\tilde{\varphi}_{iB}$ and $\tilde{\varphi}_{iC}$ respectively defined in (5.36), (5.37), (5.38), (5.40) and (5.41). The expression (5.42) can be formally introduced for all the finite volume cells; indeed, if $\mathcal{S}_{iX} = \{\emptyset\}$ ($X \in \{I, O, B, C\}$) then $\|\mathbf{v}_{iX}\| = 0$ and the term $\tilde{\varphi}_{iX}$ correctly vanishes. As a result, by combining (5.2) and (5.42), the following semi-discrete formulation of the considered balance (2.3) is finally obtained:

$$\mu_i \frac{d}{dt} \mathbf{q}_i + \tilde{\varphi}_i = \mathbf{0} \quad , \quad i \in \mathcal{I}. \tag{5.43}$$

5.1.5. Extension to rotating frames

Let $\mathcal{B}$ denote a rigid body immersed within the flow, which rotates with constant angular velocity ω (e.g. an axial inducer of the type of those introduced in Chapter 1). The representation of the governing equations with respect to a frame rotating with $\mathcal{B}$ (hereafter referred to as body-frame) is given in (2.12); the corresponding semi-discrete formulation is considered in the present section.

The external portion $\partial\mathcal{D}^{\mathrm{pol(ext)}}$ of the boundary $\partial\mathcal{D}^{\mathrm{pol}}$ in (5.33) is clearly given by:

$$\partial\mathcal{D}^{\mathrm{pol(ext)}} := \partial\mathcal{D}^I \cup \partial\mathcal{D}^O \cup \partial\mathcal{D}^C .$$

The surface $\partial\mathcal{D}^{\mathrm{pol(ext)}}$ is here assumed to be symmetrical with respect to the rotation axis; in such a circumstance, it behaves like a fixed one in the body-frame and therefore it is possible to discretize the balance (2.12) without dealing with moving computational grids. While the previous assertion is clear as far as the inflow and outflow components are concerned,[4] it may be useful to further discuss the term related to the external wall $\partial\mathcal{D}^C$. In the body frame, $\partial\mathcal{D}^C$ is a moving surface on which the slip condition is properly formulated as follows [88]:

$$\mathbf{u}^T \cdot \hat{\mathbf{n}} = (\boldsymbol{\omega} \wedge \mathbf{x})^T \cdot \hat{\mathbf{n}} \qquad (5.44)$$

where the vector product on the right-hand side represents the dragging velocity associated with the point on $\partial\mathcal{D}^C$ which is identified by the position vector $\mathbf{x}$. However, by virtue of the assumed symmetry, the vectors $\boldsymbol{\omega}$, $\mathbf{x}$ and $\hat{\mathbf{n}}$ are necessarily coplanar and therefore the right-hand side of (5.44) is systematically equal to zero. As a result, the condition (5.44) reduces to its non-rotating counterpart (5.39) and $\partial\mathcal{D}^C$ behaves as a non-rotating boundary.

In view of the aforementioned considerations, it is possible to derive the sought semi-discrete formulation from the non-rotating one (5.43), as described below. Let $\mathbf{g}_i$ denote the centroid associated with C_i, namely:

$$\mathbf{g}_i := \frac{1}{\mu_i} \int_{C_i} \mathbf{x}\, dV . \qquad (5.45)$$

Moreover, let $\mathbf{r}_i$ denote the vector mapping the projection of $\mathbf{g}_i$ on the rotation axis to $\mathbf{g}_i$ itself:

$$\mathbf{r}_i := -\hat{\boldsymbol{\omega}} \wedge \left(\hat{\boldsymbol{\omega}} \wedge \mathbf{g}_i\right)$$

where $\hat{\boldsymbol{\omega}}$ represents the versor associated with $\boldsymbol{\omega}$. Then, once introduced the following definition (derived from (2.13)):

$$\mathbf{s}_i := \|\boldsymbol{\omega}\| \begin{pmatrix} 0 \\ -2\,\hat{\boldsymbol{\omega}} \wedge \rho_i \mathbf{u}_i + \rho_i \,\|\boldsymbol{\omega}\|\, \mathbf{r}_i \end{pmatrix} \qquad (5.46)$$

---

[4] Of course, the rotation affects the representation of the fictitious state vectors $\mathbf{q}_i^I$ and $\mathbf{q}_i^O$ appearing in the approximations (5.37) and (5.38) which, however, can be formally kept.

it is possible to approximate the right-hand side of the balance (2.12) -written for C_i- as follows:

$$\int_{C_i} \mathbf{s} \, dV \approx \mu_i \, \mathbf{s}_i . \tag{5.47}$$

Finally, by combining (5.43) and (5.47), it is straightforward to introduce the following semi-discrete formulation for the considered balance (2.12):

$$\mu_i \frac{d}{dt} \mathbf{q}_i + \tilde{\varphi}_i = \mu_i \, \mathbf{s}_i \quad , \quad i \in \mathcal{I} \tag{5.48}$$

5.2. Time discretization

A discrete scheme is presented, based on a generalization of the linearized implicit time-advancing proposed in Section 3.5.

5.2.1. Linearization of the 3D Roe numerical flux

It turns out to be straightforward to extend the proposed linearization (3.137) of the preconditioned, augmented-1D Roe flux function to the 3D case. Indeed, as highlighted in Section 5.1.2, the preconditioner $\mathbf{P}_{ij}$ defined in (5.27) and the Roe matrix $\tilde{\mathbf{J}}_{ij}$ defined in (5.18) respectively generalize their augmented-1D counterparts (namely $\mathbf{P}_{ij}^{(A)}$ and $s_{ij} \tilde{\mathbf{J}}_{ij}^{(A)}$) and therefore the linearization of the preconditioned Roe numerical flux (5.30) reads (δ^n being defined in (3.11)):

$$\delta^n \boldsymbol{\phi}^P \left(\mathbf{q}_i, \mathbf{q}_j, \hat{\mathbf{v}}_{ij} \right) \approx \mathbf{A}_{ij}^n \cdot \delta^n \mathbf{q}_i + \mathbf{B}_{ij}^n \cdot \delta^n \mathbf{q}_j \tag{5.49}$$

where:

$$\begin{cases} \mathbf{A}_{ij}^n := \mathbf{A} \left(\mathbf{q}_i^n, \mathbf{q}_j^n, \hat{\mathbf{v}}_{ij} \right) \\[2mm] \mathbf{B}_{ij}^n := \mathbf{B} \left(\mathbf{q}_i^n, \mathbf{q}_j^n, \hat{\mathbf{v}}_{ij} \right) \end{cases} \tag{5.50}$$

$$\begin{cases} \mathbf{A} \left(\mathbf{q}_i, \mathbf{q}_j, \hat{\mathbf{v}}_{ij} \right) := \left(\mathbf{P}_{ij} \right)^{-1} \cdot \left(\mathbf{P}_{ij} \cdot \tilde{\mathbf{J}}_{ij} \right)^+ \\[2mm] \mathbf{B} \left(\mathbf{q}_i, \mathbf{q}_j, \hat{\mathbf{v}}_{ij} \right) := \left(\mathbf{P}_{ij} \right)^{-1} \cdot \left(\mathbf{P}_{ij} \cdot \tilde{\mathbf{J}}_{ij} \right)^- . \end{cases} \tag{5.51}$$

5.2.2. Linearized implicit time-advancing

Starting from the semi-discrete formulation (5.48), a linearized implicit time-advancing strategy is defined, as described below:

- the time derivative term in (5.48) is approximated by a backward finite difference, namely:

$$\mu_i \frac{d}{dt} \mathbf{q}_i \approx \frac{\mu_i}{\delta^n t} \delta^n \mathbf{q}_i = \frac{\mu_i}{\delta^n t} \mathbf{I} \cdot \delta^n \mathbf{q}_i \qquad (5.52)$$

- from the relevant definition (5.42), the variation of the term $\tilde{\boldsymbol{\varphi}}_i$ in (5.48) reads:

$$\delta^n \tilde{\boldsymbol{\varphi}}_i = \sum_{j \in s(i)} \delta^n \tilde{\boldsymbol{\varphi}}_{ij} + \delta^n \tilde{\boldsymbol{\varphi}}_{iI} + \delta^n \tilde{\boldsymbol{\varphi}}_{iO} + \delta^n \tilde{\boldsymbol{\varphi}}_{iB} + \delta^n \tilde{\boldsymbol{\varphi}}_{iC}. \qquad (5.53)$$

Then, in consideration of the definitions (5.36)-(5.38) and by recalling the material discussed in Section 5.2.1, it is possible to introduce the following approximations:

$$\delta^n \tilde{\boldsymbol{\varphi}}_{ij} \approx \|\boldsymbol{v}_{ij}\| \left(\mathbf{A}_{ij}^n \cdot \delta^n \mathbf{q}_i + \mathbf{B}_{ij}^n \cdot \delta^n \mathbf{q}_j \right) \qquad (5.54)$$

$$\delta^n \tilde{\boldsymbol{\varphi}}_{iI} \approx \|\boldsymbol{v}_{iI}\| \left(\mathbf{A}_{iI}^n \cdot \delta^n \mathbf{q}_i + \mathbf{B}_{iI}^n \cdot \delta^n \mathbf{q}_i^I \right)$$

$$\delta^n \tilde{\boldsymbol{\varphi}}_{iO} \approx \|\boldsymbol{v}_{iO}\| \left(\mathbf{A}_{iO}^n \cdot \delta^n \mathbf{q}_i + \mathbf{B}_{iO}^n \cdot \delta^n \mathbf{q}_i^O \right) \qquad (5.55)$$

where $\mathbf{A}_{ij}^n$ and $\mathbf{B}_{ij}^n$ are given by (5.50) and the remaining coefficients are defined in the spirit of (5.50), by suitably replacing $\mathbf{q}_j$ with the fictitious state vectors $\mathbf{q}_i^I$ and $\mathbf{q}_i^O$. Moreover, let $\mathbf{K}^{(\hat{v})}(\mathbf{q})$ denote the following Jacobian:

$$\mathbf{K}^{(\hat{v})}(\mathbf{q}) := \partial_{\mathbf{q}} \begin{pmatrix} 0 \\ p\,\hat{\boldsymbol{v}} \end{pmatrix} = \begin{pmatrix} 0 & \mathbf{0}^T \\ a^2\,\hat{\boldsymbol{v}} & \mathbf{O} \end{pmatrix}.$$

Then, by defining the following matrices:

$$\mathbf{K}_{iB}^n := \mathbf{K}^{(\hat{v}_{iB})}\left(\mathbf{q}_i^n\right) \quad , \quad \mathbf{K}_{iC}^n := \mathbf{K}^{(\hat{v}_{iC})}\left(\mathbf{q}_i^n\right)$$

it is possible to introduce the following linearization for the remaining numerical fluxes in (5.53):

$$\delta^n \tilde{\boldsymbol{\varphi}}_{iB} \approx \|\boldsymbol{v}_{iB}\| \, \mathbf{K}_{iB}^n \cdot \delta^n \mathbf{q}_i \qquad (5.56)$$

$$\delta^n \tilde{\boldsymbol{\varphi}}_{iC} \approx \|\boldsymbol{v}_{iC}\| \, \mathbf{K}_{iC}^n \cdot \delta^n \mathbf{q}_i. \qquad (5.57)$$

By combining (5.54)-(5.55), (5.56) and (5.57) it is possible to recast (5.53) as follows:

$$\delta^n \tilde{\boldsymbol{\varphi}}_i \approx \mathbf{F}_{ii}^n \cdot \delta^n \mathbf{q}_i + \sum_{j \in s(i)} \mathbf{F}_{ij}^n \cdot \delta^n \mathbf{q}_j + \mathbf{F}_{iI}^n \cdot \delta^n \mathbf{q}_i^I + \mathbf{F}_{iO}^n \cdot \delta^n \mathbf{q}_i^O \quad (5.58)$$

where:

$$
\begin{cases}
\mathbf{F}_{ii}^n := \displaystyle\sum_{j \in s(i)} \| \mathbf{v}_{ij} \| \, \mathbf{A}_{ij}^n \\[2mm]
\qquad\quad + \| \mathbf{v}_{iI} \| \, \mathbf{A}_{iI}^n + \| \mathbf{v}_{iO} \| \, \mathbf{A}_{iO}^n + \| \mathbf{v}_{iB} \| \, \mathbf{K}_{iB}^n + \| \mathbf{v}_{iC} \| \, \mathbf{K}_{iC}^n \\[2mm]
\mathbf{F}_{ij}^n := \| \mathbf{v}_{ij} \| \, \mathbf{B}_{ij}^n \\[2mm]
\mathbf{F}_{iI}^n := \| \mathbf{v}_{iI} \| \, \mathbf{B}_{iI}^n \\[2mm]
\mathbf{F}_{iO}^n := \| \mathbf{v}_{iO} \| \, \mathbf{B}_{iO}^n
\end{cases}
$$

- let $\mathbf{S}_i$ denote the Jacobian of the term $\mathbf{s}_i$ introduced in (5.46), namely:

$$
\mathbf{S}_i := \partial_{\mathbf{q}_i} \partial \mathbf{s}_i = \| \boldsymbol{\omega} \|
\begin{pmatrix}
0 & \mathbf{0}^T \\[2mm]
\| \boldsymbol{\omega} \| \, \mathbf{r}_i & -2\,\boldsymbol{\Omega}
\end{pmatrix}
\tag{5.59}
$$

with $\boldsymbol{\Omega}$ defined by the following relation:

$$
\hat{\boldsymbol{\omega}} \wedge \mathbf{y} = \boldsymbol{\Omega} \cdot \mathbf{y}
$$

where $\mathbf{y}$ is a generic vector in $\mathbb{R}^3$. Once noticed that the matrix (5.59) does not depend on the specific instance of the state vector $\mathbf{q}_i$ (and therefore on the time-level), it is possible to linearize the right-hand side of (5.48) as follows:

$$
\mu_i \, \mathbf{s}_i^{n+1} = \mu_i \, \mathbf{s}_i^n + \mu_i \, \mathbf{S}_i \cdot \delta^n \mathbf{q}_i .
\tag{5.60}
$$

By combining (5.52), (5.58) and (5.60), it is straightforward to introduce the following discrete scheme:

$$
\mathbf{E}_i^n \cdot \delta^n \mathbf{q}_i + \sum_{j \in s(i)} \mathbf{F}_{ij}^n \cdot \delta^n \mathbf{q}_j + \mathbf{F}_{iI}^n \cdot \delta^n \mathbf{q}_i^I + \mathbf{F}_{iO}^n \cdot \delta^n \mathbf{q}_i^O = \mathbf{b}_i^n \quad , \quad i \in \mathcal{I}
\tag{5.61}
$$

where:

$$
\begin{cases}
\mathbf{E}_i^n := \dfrac{\mu_i}{\delta^n t} \mathbf{I} + \mathbf{F}_{ii}^n - \mu_i \, \mathbf{S}_i \\[3mm]
\mathbf{b}_i^n := \mu_i \, \mathbf{s}_i^n - \tilde{\boldsymbol{\varphi}}_i^n .
\end{cases}
$$

Note 5.2.1. The equation (5.61) clearly represents a sparse linear system which can be solved once the boundary terms $\delta^n \mathbf{q}_i^I$ and $\delta^n \mathbf{q}_i^O$ have been suitably associated to specific BCs. For instance, if a uniform inflow is assumed with respect to the non-rotating frame, associated with the state vector $\mathbf{q}_\infty(t)$, then $\mathbf{q}_i^I$ admits the following representation in the body frame:

$$\mathbf{q}_i^I = \mathbf{q}_\infty(t) - \|\boldsymbol{\omega}\| \begin{pmatrix} 0 \\ \hat{\boldsymbol{\omega}} \wedge \mathbf{g}_i \end{pmatrix} \tag{5.62}$$

with $\mathbf{g}_i$ given by (5.45). In consideration of the fact that the corresponding variation:

$$\delta^n \mathbf{q}_i^I = \mathbf{q}_\infty(t^{n+1}) - \mathbf{q}_\infty(t^n)$$

does not involve any unknown, the term $\mathbf{F}_{iI}^n \cdot \delta^n \mathbf{q}_i^I$ in (5.61) must be formally incorporated into the known term $\mathbf{b}_i^n$. If, in addition, the following transmissive outflow BC is assumed:

$$\mathbf{q}_i^O = \mathbf{q}_i \tag{5.63}$$

then the corresponding variation, namely:

$$\delta^n \mathbf{q}_i^O = \delta^n \mathbf{q}_i$$

clearly implies that the coefficient $\mathbf{F}_{iO}^n$ in (5.61) must be formally incorporated into $\mathbf{E}_i^n$. As a result, when adopting the BCs (5.62) and (5.63), the system (5.61) becomes:

$$\bar{\mathbf{E}}_i^n \cdot \delta^n \mathbf{q}_i + \sum_{j \in s(i)} \mathbf{F}_{ij}^n \cdot \delta^n \mathbf{q}_j = \bar{\mathbf{b}}_i^n \tag{5.64}$$

with:

$$\begin{cases} \bar{\mathbf{E}}_i^n := \dfrac{\mu_i}{\delta^n t} \mathbf{I} + \mathbf{F}_{ii}^n + \mathbf{F}_{iO}^n - \mu_i \, \mathbf{S}_i \\[2mm] \bar{\mathbf{b}}_i^n := \mu_i \, \mathbf{s}_i^n - \tilde{\boldsymbol{\varphi}}_i^n - \mathbf{F}_{iI}^n \cdot \delta^n \mathbf{q}_i^I \, . \end{cases}$$

Chapter 6
3D Applications

In the present section, the numerical method proposed in Chapter 5 is applied to the liquid flow around a hydrofoil (Section 6.1) as well as to the flow around an axial inducer (Section 6.2). For both cases, suitable instances of the barotropic state law introduced in Section 4.1 are adopted.

6.1. Simulation of the 3D flow around a hydrofoil

The water flow around a 3D NACA0015 hydrofoil having chord $c = 115$ mm and mounted within a water tunnel at 4° angle of attack is considered, as a validation benchmark for the linearized implicit scheme proposed in Chapter 5. After introducing the problem in Section 6.1.1, some issues regarding the numerical discretization as well as the used computational resource are presented in Sections 6.1.2 to 6.1.4. Non-cavitating as well as cavitating numerical simulations are respectively presented in Sections 6.1.5 and 6.1.6.

6.1.1. Problem description

The geometry of the test-chamber is sketched in Figure 6.1 while the test-section, which is obtained by cutting the chamber along its symmetry plane, is sketched in Figure 6.2.

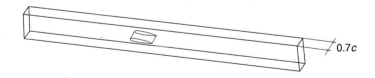

Figure 6.1. Sketch of the 3D test-chamber.

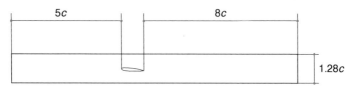

Figure 6.2. Sketch of the test-section.

The considered temperature of the water is $T_L = 293.16$ K. Let the subscript ∞ denote the free-stream (unperturbed) conditions; experimental data are available for the conditions reported in Table 6.1.

Table 6.1. Free-stream conditions of the available experiments.

Free-stream	p_∞ (Pa)	$\|\mathbf{u}_\infty\|$ (m/s)	M_∞	σ_∞
FS1	59050	3.115	$2.2 \cdot 10^{-3}$	11.7
FS2	12000	3.460	$2.4 \cdot 10^{-3}$	1.5

More precisely, measurements of the pressure coefficient:

$$C_p := \frac{p - p_\infty}{\frac{1}{2} \rho_\infty \|\mathbf{u}_\infty\|^2}$$

are available [81] along the curve which is defined by intersecting the hydrofoil surface and the test-section. The velocity $\mathbf{u}_\infty$ is orthogonal to the inlet section. The symbols M_∞ and σ_∞ in Table 6.1 respectively denote the free-stream Mach number, defined as follows:

$$M_\infty := \frac{\|\mathbf{u}_\infty\|}{a_\infty} \tag{6.1}$$

and the cavitation number, defined as follows [9]:

$$\sigma_\infty := \frac{p_\infty - p_{sat}}{\frac{1}{2} \rho_\infty \|\mathbf{u}_\infty\|^2} . \tag{6.2}$$

In view of the definition (6.2), it is clear that cavitation phenomena are likely to take place in correspondence of low cavitation numbers. The conditions FS1 in Table 6.1, in particular, are associated with a non-cavitating flow, which can be considered as a (very) low Mach number validation benchmark for numerical solvers. Conversely, the conditions FS2 are associated with a cavitating flow. At the considered liquid temperature, the transition between non-cavitating and cavitating flow regions is extremely abrupt [10]; this behaviour is described by e.g. the complex state laws shown in Figures 4.1 and 4.2, whose numerical treatment is particularly tough.

6.1.2. Computational grids

The domain sketched in Figure 6.1 is discretized by means of a 3D tetra-hedral unstructured grid. The considered grids are reported in Table 6.2, in which the symbols N_c and N_t (defined in Section 5.1.1) respectively denote the number of cells (*i.e.* nodes) and elements (*i.e.* tetrahedra). It

Table 6.2. Considered computational grids.

Grid	N_c	N_t
GR1	27220	137756
GR2	19322	88400

is worth mentioning that:

- both the grids GR1 and GR2 are 3D tetrahedral, unstructured grids. However, by construction they are symmetrical with respect to the test-section and therefore their imprint on the test-section appears as a 2D triangular, unstructured grid (see Figure 6.3);
- neither GR1 nor GR2 is highly refined in order to contain the computational cost of the simulations while validating/developing the considered numerical schemes (examples of finer grids discretizing the domain under consideration can be found in [6]);
- while GR1 represents the whole test-chamber, GR2 only discretizes a "slice" of it (its span-wise width being 0.1 c instead of 0.7 c) and it is used for reducing the computational cost while validating/developing the considered numerical schemes.

The considered grids must be partitioned in order to be incorporated into the parallel numerical frame mentioned in the introduction to the present document (*i.e.* the AERO code). The grids GR1 and GR2, in particular, have been divided into 5 and 2 sub-domains, respectively. To the purpose, the proprietary software "TopDomDec" as well as the open source tool "Metis" (http://www-users.cs.umn.edu/ karypis/metis/metis) have been exploited.

6.1.3. Computational resources

The considered computational resources are reported in Table 6.3. Among them, COMP3 denotes the IBM SP4 computing platform available at CINECA (currently upgraded to SP5, see http://www.cineca.it) while COMP1 and COMP2 are common PCs.

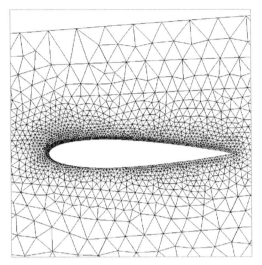

Figure 6.3. Imprint of the grid GR1 on the test-section (detail).

Table 6.3. Available computers.

Computer	CPU	No. of CPUs	Total RAM
COMP1	Intel Pentium4, 2.66 GHz	1	512 MB
COMP2	Intel Pentium4 Xeon, 3.06 GHz	2	8 GB
COMP3	IBM POWER4, 1.3 GHz	512	1088 GB

6.1.4. Numerical discretization

The linearized implicit discrete scheme which is derived from (5.64) by setting $\omega = 0$ is considered for both the non-cavitating and the cavitating simulations.

A variable time-step is adopted, defined as follows:

$$\delta^n t = c^{(\text{CFL})n} \min_{h \in \mathcal{H}} \left(\frac{\lambda_h}{\tilde{s}_h^n} \right) \tag{6.3}$$

where:

- λ_h denotes the minimum among the four heights which are associated with the $h-$th tetrahedron T_h ($h \in \mathcal{H} := \{1, \ldots, N_t\}$);
- $\tilde{s}_h^n$ denotes the value at time-level n of an estimate of the maximum wave speed $\tilde{s}_h$ associated with T_h. More precisely, $\tilde{s}_h$ is chosen as the maximum among the wave speeds arising in the Roe-linearized RPs associated with the four vertices of T_h;

- $c^{(CFL)n}$ denotes the value at time-level n of a CFL-like coefficient, $c^{(CFL)}$, which can be modulated during the simulation (see below).

Since, as shown by the experiments, the flows associated with the considered free-streams in Table 6.1 turn out to be substantially steady (even the cavitating one, due to the low angle of attack), the numerical simulations are advanced in time up to a steady-state.

6.1.5. Non-cavitating simulations

The considered non-cavitating test-cases are summarized in Table 6.4. The following state vector:

$$\mathbf{q}_\infty = \begin{pmatrix} \rho_\infty \\ \rho_\infty \, \mathbf{u}_\infty \end{pmatrix} \qquad (6.4)$$

is derived, in particular, from the chosen free-stream FS1 (see Table 6.1 above). The state vector (6.4), in turn, is introduced in (5.62) for defining the fictitious inflow state vectors $\mathbf{q}_i^I$; moreover, it is exploited for defining the adopted initial conditions. The free-stream Mach number M_∞ (see Table 6.1 above) is assumed to be the characteristic Mach number $M_\star$ to be used for preconditioning the Roe numerical flux; in particular, the constant β_{ref} in (3.97) is chosen equal to 1.[1] Furthermore, the parameters "Liquid" and "T_L" in Table 6.4 characterize the isentropic compressibility coefficient $\vartheta \approx 8.55 \cdot 10^5$ appearing in the chosen liquid model (4.1).

Table 6.4. Considered non-cavitating test-cases.

Test-case	Free-stream	Liquid	T_L (K)	Grid
NONCAV1	FS1	water	293.16	GR1
NONCAV2	FS1	water	293.16	GR2

The resulting C_p distribution is reported in Figure 6.4, against the relevant experimental data. It is worth noticing that:

---

[1] A sensitivity study has been performed in order to set β_{ref}, not reported here for the sake of conciseness. It has been observed that smaller time-steps must be adopted when decreasing β_{ref} from its upper bound, $\beta_{ref} = M_\star^{-1}$ (corresponding to the non-preconditioned case $\beta = 1$ in (3.97)) down to its "recommended" value, i.e. $O(1)$. However, the resulting numerical solution turns out to be considerably inaccurate (in terms of C_p, against the experiments) when β_{ref} is distant from 1; vice versa, for β_{ref} in the neighbourhood of the unity, the most accurate result seems to be associated with $\beta_{ref} \approx 1$.

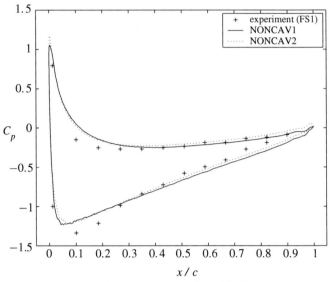

Figure 6.4. C_p distribution for the test-cases in Table 6.4.

- the numerical results respectively obtained by exploiting GR1 and GR2 are close to each other, thus showing that the obtained C_p distribution is almost independent of the grid. Moreover, no appreciable 3D effects take place along the span-wise direction (which is not surprising, by virtue of the assumed absence of viscosity effects);
- the agreement between the numerical results and the experimental data can be considered reasonably good, in view of the fact that the considered 3D numerical scheme is only first-order accurate and the used grids are relatively coarse. Both these issues seem to contribute, for instance, to underestimating the suction peak which is located near the leading edge of the hydrofoil;[2]
- the coefficient $c^{(CFL)}$ in (6.3) has been increased, linearly with respect to n, during the first iterations for smoothly abandoning the initial flow field (which is, in general, a crude approximation of the final one). In particular, it has been increased up to more than 400 for both the considered test-cases, thus confirming the efficiency of the proposed linearized implicit schemes when dealing with smooth flows (see Section 3.5.5). As far as the total CPU time is concerned, the test-case NONCAV1 requires 17 hours and 30 minutes on the computer COMP3

---

[2] A better result could be obtained by suitably refining the grid in the leading edge area and by increasing the order of spatial accuracy of the scheme (*e.g.* by a standard MUSCL technique [108]); these improvements are postponed to a subsequent research stage.

reported in Table 6.3 (a contained elapsed time is obtained, due to the parallelization strategy) while the test-case NONCAV2 requires 7 hours and 30 minutes on the computer COMP1[3] reported in Table 6.3.

6.1.6. Cavitating simulations

The considered cavitating test-cases are reported in Table 6.5. A state vector of the type of (6.4) is introduced also for the present case, based on the free-stream FS3 defined in Table 6.6. More in detail, the considered state vector is obtained from that one associated with the free-stream FS1 in Table 6.1 (Section 6.1.1), by decreasing the pressure p_∞; such a procedure has been actually performed for defining the considered inlet and initial conditions.[4] The free-stream Mach number M_∞ in Table 6.6 is assumed to be the characteristic Mach number of the liquid region. However, in consideration of the fact that no preconditioning is required within the cavitating region (where the flow can be easily hypersonic), a local preconditioning strategy is heuristically adopted. More precisely, a local preconditioning parameter β_{ij}^2, defined as follows (compare with (3.142)):

$$\beta_{ij}^2 := \begin{cases} M_\infty^2 & \text{if} \quad \min\left(\rho_i, \rho_j\right) \geq \rho_{\text{Lsat}} \\ 1 & \text{otherwise} \end{cases}$$

is introduced into the preconditioning matrix (5.27) in place of the original parameter β^2. As far as the state law is concerned, the parameters "Liquid", "T_L" and "ζ" in Table 6.5 characterize two instances of the barotropic model (4.1)-(4.9). The relevant model parameters are $\vartheta \approx 8.55 \cdot 10^5$, $\sigma_1 \approx 1.33 \cdot 10^3$, $\sigma_2 \approx -0.73$ and $\sigma_3 \approx 0.78$; the mixture branches of the considered laws are shown in Figure 4.1.

Table 6.5. Considered cavitating test-cases.

Test-case	Free-stream	Liquid	T_L (K)	ζ	Grid
CAV1	FS3	water	293.16	0.1	GR1
CAV2	FS3	water	293.16	0.1	GR2
CAV3	FS3	water	293.16	0.01	GR2

---

[3] The considered parallel code has been run on the mono-processor computer COMP1 by means of the "LAM" parallel environment (see http://www.lam-mpi.org), thus introducing a certain degree of communication overhead.

[4] A variable free-stream state vector $q_\infty(t)$ is explicitly considered in (5.62). Moreover, a user-defined flow field (typically, the result of a previous simulation) can be read by the developed numerical solver for starting the simulation.

Table 6.6. Considered free-stream conditions.

Free-stream	p_∞ (Pa)	$\|\mathbf{u}_\infty\|$ (m/s)	M_∞	σ_∞
FS3	7500	3.115	$2.2 \cdot 10^{-3}$	1.1

It is possible to adopt the experimental data based on the free-stream FS2 in Table 6.1 (Section 6.1.1) for validating the cavitating simulations at hand, since the corresponding cavitation number is similar to that one associated with the considered free-stream FS3. Hence, in Figures 6.5 and 6.6, the C_p distribution -on the suction side of the considered hydrofoil- which is obtained from the considered simulations is compared with the aforementioned experimental points.

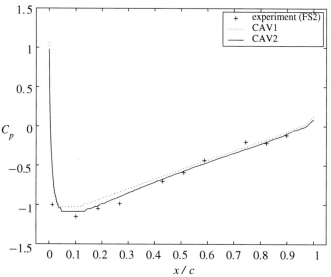

Figure 6.5. C_p distribution (suction side) for the test-cases CAV1 and CAV2 reported in Table 6.5.

It is worth noticing that:

- only small differences, located near the leading edge of the hydrofoil (*i.e.* where cavitation occurs), appear in the C_p distribution when adopting different grids (see Figure 6.5);
- the agreement between the numerical results and the experimental data can be considered reasonably good, in view of the fact that it is very challenging to accurately simulate the cavitation phenomena at hand;

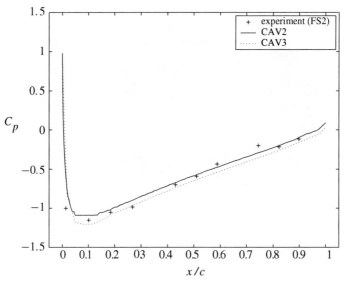

Figure 6.6. C_p distribution (suction side) for the test-cases CAV2 and CAV3 reported in Table 6.5.

- on the basis of the numerical results in Figure 6.6, it seems that the C_p distribution gradually varies with respect to ζ. Moreover, a lower value of ζ correctly leads to a less pronounced Mach number variation (indeed, as shown in Figure 4.2, the minimum sound speed -in the cavitating region- increases when reducing ζ), as shown in Figures 6.7 and 6.8.

  Furthermore, once defined a local cavitation number as follows (compare with (6.2)):

  $$\sigma := \frac{p - p_{\text{sat}}}{\frac{1}{2}\,\rho_\infty\,\|\mathbf{u}_\infty\|^2}$$

  it is possible to identify the cavity with the fluid sub-domain within which $\sigma < 0$. Then, as shown in Figures 6.9 and 6.10, it is possible to see that a lower value of ζ results in a more extended cavity. Also this result seems to be correct, since the nearly constant pressure value in Figure 4.1, which roughly provides a characteristic value of the cavity pressure, decreases when decreasing ζ. Nevertheless, a systematic investigation of the sensitivity of the numerical results to the free cavitation model parameter ζ is postponed to a further research stage;

- before the inception of cavitation, the coefficient $c^{(\text{CFL})}$ introduced in (6.3) can be increased during the simulation up to $O\left(10^2\right)$ for all the considered cavitating test-cases However, as soon as cavitation oc-

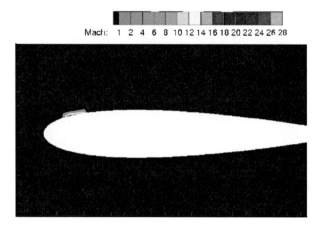

Figure 6.7. Contour plot of the local Mach number on the test-section (detail), for the test-case CAV2 reported in Table 6.5.

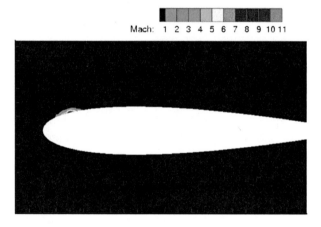

Figure 6.8. Contour plot of the local Mach number on the test-section (detail) for the test-case CAV3 reported in Table 6.5.

curs, it must be reduced to $O\left(10^{-2}\right)$ for all the considered simulations to remain stable. This point seems to confirm the hypothesis put forward in Section 3.5.6 according to which the observed stability restriction can be caused by the presence/onset of discontinuities in the flow field (caused by the cavitation inception in the present case), which render it more difficult to exploit the proposed linearized scheme. As far as the total CPU time is concerned, the test-case CAV1 approximately requires 400 hours on the computer COMP3 (see Table 6.3) while the test-cases CAV2 and CAV3 approximately require 150 hours

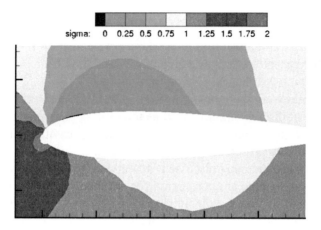

Figure 6.9. Contour plot of the local cavitation number (sigma) on the test-section (detail), for the test-case CAV2 reported in Table 6.5.

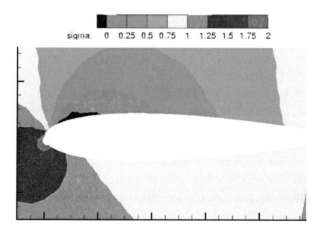

Figure 6.10. Contour plot of the local cavitation number (sigma) on the test-section (detail), for the test-case CAV3 reported in Table 6.5.

on the computer COMP2. In both cases, a contained elapsed time is obtained, thanks to the parallelization strategy.

Further investigation is definitely recommended in order to counteract the aforementioned efficiency problem. Moreover, according to the author, it would of interest to also assess the effects that the chosen local pre-conditioning strategy produces on the stability properties of the resulting numerical scheme. However, such a study is postponed to a subsequent research stage.

6.2. Simulation of the 3D flow around an axial inducer

The water flow around a turbo-pump inducer (see Section 1.1) is considered in the present section, as a qualitative validation benchmark for the linearized implicit scheme proposed in Chapter 5.

After introducing the problem in Section 6.2.1, some issues regarding the numerical discretization as well as the used computational resource are presented in Sections 6.2.2 to 6.2.4. Some non-cavitating numerical results are finally presented in Section 6.2.5. In consideration of the efficiency problems already discussed in Sections 3.5.6, 4.2 and 6.1.6, no cavitating simulations are considered for the inducer flow at hand. Indeed, the huge increase in computational cost, which is here amplified by the complexity of the considered geometry (see below), makes it practically impossible to advance the simulation unless exploiting specific supercomputing resources, that are not available within the scope of the present research project.[5]

6.2.1. Problem description

The considered geometry is sketched in Figure 6.11, where the inducer is denoted by "I". A nose "N" as well as an after-body "A" smoothly join "I"; in particular, the nose is part of an axisymmetrical ellipsoid while the after-body is a circular cylinder having a diameter equal to the base diameter of the inducer. The flow domain is bounded by a cylindrical case, whose diameter is equal to the maximum blade tip diameter D; hence, there is no tip clearance and a shrouded inducer (see Section 1.1) is considered. The length L_{out} of the after-body, as well as the length L_{in} of the inflow section, are chosen equal to 1.5 D. The inducer angular speed is equal to 2000 rpm.

The chosen temperature of the water is $T_L = 296.16$ K. The considered free-stream conditions are reported in Table 6.7.

Table 6.7. Considered free-stream conditions.

Free-stream	p_∞ (Pa)	$\|\mathbf{u}_\infty\|$ (m/s)	M_∞	σ_∞
FS4	115000	0.476	$3.4 \cdot 10^{-4}$	990

Both M_∞ and σ_∞ in the aforementioned table, respectively defined in (6.1) and (6.2), are computed by exploiting the absolute velocity $\mathbf{u}_\infty$;

---

[5] As mentioned in [93], a cavitating simulation has been stopped at the inception stage, due to the aforementioned efficiency problems.

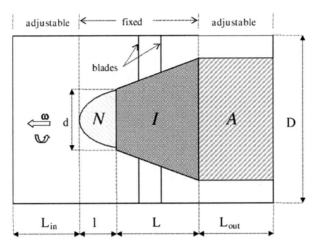

Figure 6.11. Schematic representation of the considered inducer geometry.

the local Mach number and the local cavitation number are respectively higher and lower than those reported in Table 6.7, due to the dragging velocity appearing in the body frame.

6.2.2. Computational grids

The domain sketched in Figure 6.11 is discretized by means of a 3D tetrahedral unstructured grid, whose main features are reported in Table 6.8 (N_c and N_t respectively denoting the number of nodes and elements). It is worth mentioning that:

Table 6.8. Considered computational grid.

Grid	N_c	N_t
GR3	549139	2588501

- the size of the grid elements smoothly transitions between different regions on the body surface (*e.g.* the nose-inducer junction shown in Figure 6.12) and accurately follows the solid walls even within high-curvature regions (*e.g.* the hub-blade intersection shown in Figure 6.13);
- as far as the external case is concerned, it is not possible to define a perfectly cylindrical wall due to the numerical errors (even if very small) related to the numerical format of the inducer geometry file. To counteract this problem, a kind of shell covering the inter-blade passages is modelled, whose external aspect is shown in Figure 6.14.

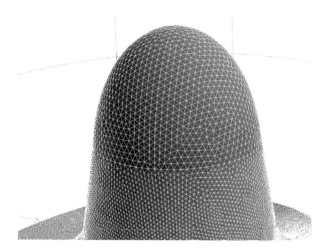

Figure 6.12. Detail of the grid GR3 at the nose-inducer junction.

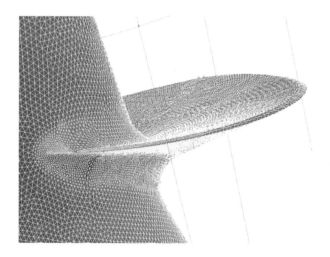

Figure 6.13. Detail of the hub-blade intersection for GR3.

The considered grid has been partitioned into 16 sub-domains in order to be incorporated into the parallel numerical frame mentioned in the introduction to the present document (*i.e.* the AERO code). To the purpose, the proprietary software "TopDomDec" has been exploited.

6.2.3. Computational resources

In consideration of the noticeable size of the grid at hand, the only super-computer COMP3 reported in Table 6.3 (Section 6.1.3) is considered.

Figure 6.14. External view of the inter-blade covering created for GR3. The "cut" on the boundary surface represent the imprint of the inducer blade tip.

6.2.4. Numerical discretization

The linearized implicit discrete scheme (5.64) is considered, in which the terms related to the rotation are computed by exploiting the inducer (constant) angular velocity.

As far as the time-advancing is concerned, the variable time-step (6.3) is adopted.

6.2.5. Non-cavitating simulations

The considered non-cavitating test-case is reported in Table 6.9. A state vector of the type of (6.4), derived from the considered free-stream FS4 (see Table 6.7 in Section 6.2.1), is introduced for the defining the fictitious inflow state vector $\mathbf{q}_i^I$ in (5.62). Moreover, a uniform initial flow field, determined by the free-stream conditions, is assumed; its representation in the body frame is therefore obtained (for the i–th finite volume cell) by evaluating the right-hand side of (5.62) in correspondence of the aforementioned free-stream state vector. In consideration of the fact that, with respect to the rotating frame, the local Mach number can undergo substantial variations along the radial direction due to the dragging velocity, a local preconditioning strategy is required (see the relevant paragraph in Section 3.5.4). In particular, a local preconditioning parameter β_{ij}^2 (to be introduced into the preconditioner (5.27) in place of the original parameter β^2) is heuristically defined as follows (compare with (3.142)):

$$\beta_{ij}^2 := 1 - \exp\left(-\left(\hat{M}_{ij}\right)^2\right)$$

where:

$$\hat{M}_{ij} := \frac{\|\mathbf{u}_{ij}\|}{a_{ij}}$$

with $\mathbf{u}_{ij}$ and a_{ij} respectively defined in (5.8) and (3.62). Finally, as far as the state law is concerned, the parameters "Liquid" and "T_L" in Table 6.9 characterize the isentropic compressibility coefficient $\vartheta \approx 7.13 \cdot 10^5$ appearing in the chosen liquid model (4.1).

Table 6.9. Considered non-cavitating test-case.

Test-case	Free-stream	Liquid	T_L (K)	Grid
NONCAV3	FS4	water	296.16	GR3

The pressure contours on the inducer surface, obtained after 27000 iterations, are reported in Figure 6.15.

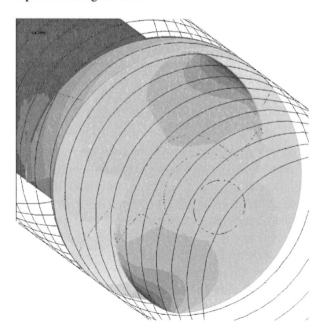

Figure 6.15. Pressure contours on the inducer surface for the test-case NON-CAV3: max [red] 177700 (Pa), min [blue] 79700 (Pa), spacing 5000 (Pa).

It is worth noticing that:

- the behaviour of the flow field, as described by the numerical simulation, is in a good qualitative agreement with that one observed in a number of experimental works. Indeed, the working fluid gradually undergoes a pressure rise while flowing within the vanes between the rotating blades, as shown in Figure 6.15. Moreover, according to this figure, the flow region which is most prone to cavitation is located near that portion of the blades where the volutes, detaching from the hub, firstly reach the external tip diameter D. This is in agreement with the experiments which, for similar flow conditions, observe the cavitation inception exactly in the flow region under consideration (see e.g. [14]);
- the considerable axial back-flow occurring near the blade tip where the diameter is less than D (i.e. where the volutes are not completely shrouded), which is well documented in a number of experimental works (e.g. [118]), is described by the numerical simulation as well, as shown in Figure 6.16;

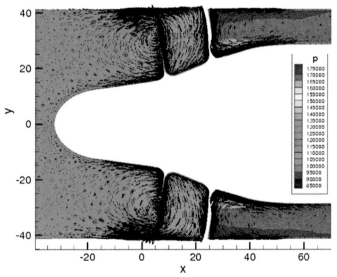

Figure 6.16. Velocity field (x: axial component, y: radial component) in a longitudinal cut plane of the flow domain for the test-case NONCAV3. Pressure contours are drawn in the background.

- the coefficient $c^{(\text{CFL})}$ introduced in (6.3) has been increased during the simulation for smoothly abandoning the initial flow field (which is, in general, a crude approximation of the final one). In particular, it has been increased up to 350. As far as the total CPU time is concerned, the considered simulation approximately requires 1500 hours on the computer mentioned in Section 6.2.3; the corresponding elapsed time can be contained by virtue of the parallelization strategy.

Chapter 7
Concluding remarks

A numerical method for simulating 3D barotropic flows in complex, possibly rotating, geometries has been presented. The considered method can successfully cope with nearly-incompressible flows by *ad hoc* preconditioning and allows for an efficient linearized implicit time-advancing technique to be defined. All the proposed numerical ingredients were implemented within a parallel numerical framework; the resulting CFD solver was validated against 3D non-cavitating as well as cavitating liquid flows. The documented research activities were driven by an industrial program, funded by the Italian Space Agency (ASI), aimed at developing a numerical tool for simulating propellant flows around 3D rotating axial inducers belonging to the feed turbo-pump system of a liquid propellant rocket engine.

In view of the fact that, under typical operational conditions, cavitation phenomena can take place within the aforementioned turbo-machines, the choice of a suitable cavitation model was initially addressed. A literature review suggested considering an equivalent fluid cavitation model; a barotropic homogeneous flow model was adopted, in particular, which can take into account thermal cavitation effects and, possibly, the concentration of the active cavitation nuclei. This model was coupled with the mass and momentum balances of classical fluid dynamics; the effects of viscosity were neglected. In order to incorporate the chosen model into an existing numerical frame which was available to the research group, namely the AERO code described in the Introduction, a density-based numerical approach was chosen. The AERO code was originally conceived for dealing with ideal gases and the specific expression and properties of the ideal gas state law deeply affected its implementation. In particular, both the definition of the Roe numerical flux function (characterizing the space discretization of the convective fluxes by a finite volume approach) and the linearized implicit time-advancing strategy (involving an approximate linearization of the aforementioned numerical flux) were based on the ideal gas state law. Moreover, also the preconditioning technique introduced for coping with low Mach number flows was affected by the specific form of the adopted state law. As a consequence, all these numerical issues needed to be replaced, if possible, with proper counterparts holding for a barotropic state law.

The definition of the new numerical ingredients was initially conceived in a 1D context; moreover, in order to keep a certain degree of generality, a generic barotropic state law was assumed. Once defined a Roe numerical flux applicable to generic barotropic fluids, the accuracy of the resulting semi-discrete formulation, as applied to nearly-incompressible flows, was addressed following [42]. This study showed that for low Mach number flows the accuracy of the proposed semi-discrete formulation degrades; the same result had already been found -and a suitable remedy (preconditioning) had been proposed- in [42] for the ideal gas case. The considered preconditioning strategy was successfully extended to the barotropic case; however, the introduction of the preconditioning narrowed the stability region of common explicit time-advancing schemes. To counteract this problem, a linearized implicit time-advancing strategy was proposed, only relying on the algebraic properties of the Roe flux function and therefore applicable to a variety of problems. In particular, differently from the linearization technique already implemented in AERO, the proposed one does not rely on the first-order homogeneity of the analytical flux function (since this properties, satisfied by the ideal gas state law, does not hold for the barotropic one). The implicit scheme was further extended so as to incorporate the aforementioned preconditioning strategy. All these ingredients were qualitatively validated in a 1D context, namely the water flow in a convergent-divergent nozzle, for both non-cavitating and cavitating conditions [91]. The proposed preconditioning technique turned out to effectively counteract the accuracy problem at low Mach numbers. Furthermore, the proposed linearized implicit scheme allowed for an efficient time-advancing to be performed when considering non-cavitating flows; as soon as cavitation occurred, however, significantly smaller time-steps had to be adopted. The proposed 1D numerical techniques were then extended to the 3D case, firstly to non-rotating and then to rotating frames. The generalization of the Roe numerical flux, in particular, was accomplished by exploiting the tensorial character of the considered governing equations while the extension to rotating frames was performed by introducing a suitable term in the aforementioned equations, accounting for non-inertial effects. The proposed 3D numerical method was firstly validated by considering the water flow around a NACA0015 hydrofoil, for which experimental data concerning the pressure coefficient distribution were available. In particular, water at

20° C was considered, possibly leading to the occurrence of "cold cavitation" phenomena whose numerical treatment is extremely challenging. All the issues highlighted in the 1D numerical experiments appeared in the 3D case as well; in particular, the proposed scheme proved out to efficiently compute non-cavitating flows but, as soon as cavitation takes place, its efficiency was significantly reduced. As far as the accuracy is concerned, the obtained results (which appeared to be independent of the grid) seemed reasonably good for both non-cavitating and cavitating conditions, in view of the fact that the considered 3D numerical scheme was only first-order accurate and the used grids were relatively coarse. A few thousand iterations of a non-cavitating simulation of the water flow around an axial turbo-pump inducer were carried out as well. The behaviour of the flow field, as described by the considered numerical simulation, turned out to be in a good qualitative agreement with that one observed in a number of experimental works. In particular, the numerical solution correctly described the pressure contours on the surface of the inducer blades as well as the considerable axial back-flow occurring where the inducer volutes are not completely shrouded. Moreover, also for the non-cavitating case under consideration, an efficient time-advancing could be performed.

A more systematic investigation of the aforementioned 1D numerical ingredients was then started. In this context, the exact solution of the 1D Riemann problem associated with a generic convex barotropic state law was addressed and a solution procedure was proposed (which was also exploited for defining exact benchmarks for the validation of the 1D numerical schemes considered in the present document). A Godunov numerical flux function based on the aforementioned exact solution was defined as well.

Clearly, the efficiency problem emerging when considering non-smooth flow fields, like those originating from cavitation inception when adopting realistic homogeneous flow models, deserves special attention. In view of the numerical results reported in the present document, this efficiency issue seems to be imputable to the approximate linearization of the Roe numerical flux in the implicit time-advancing (as briefly mentioned, the specifically adopted linearization does not seem to play a crucial role in this problem). Consequently, it could be of interest to also consider different (*i.e.* more robust, even if less refined) numerical flux functions as, for instance, the Rusanov flux, the HLL/HLLC flux, etc... [98] (in this spirit, the proposed solution to the 1D Riemann problem associated with convex barotropic state laws could be exploited for investigating further Godunov methods). The aforementioned point could be supported by the fact that the considered Roe flux function (as it stands, without fixes) may provide entropy-violating solutions within the transonic regime associated with cavitation inception. Furthermore, the fact that phase transition (and therefore cavitation) is a major reason in the lack of convexity of the considered state law [69] may add to the complexity of the problem, since the convexity may be important when seeking entropic solutions [3]. It is therefore evident that there is room for improvement while keeping the adopted numerical frame (*i.e.* homogeneous flow cavitation model, compressible -generally preconditioned-algorithms, finite volume space discretization, linearized implicit time-advancing); further investigation in this direction is definitely recommended. Simultaneously, it would be of interest to increase the order of accuracy of the proposed method (*e.g.* by developing the "Defect Correction" strategy briefly discussed in Section 3.5.3) as well as to investigate additional/different numerical ingredients as, for instance, relaxation techniques (see *e.g.* [3, 22] and [23]) and dual time-stepping strategies (see *e.g.* [22, 23] and [58]), which seem to improve the convergence properties of the considered algorithms.

As a concluding remark, it may be worth emphasizing that the assumed generality of the considered barotropic state law permits to apply the proposed material to several problems (*e.g.* to shallow water flows, see Note 2.5.1 in Section 2.5.1). This aspect, together with the fact that the proposed linearization of the Roe numerical flux function may be applied when considering an arbitrary state law (not necessarily a barotropic one), endow the present work with a certain degree of generality.

Appendix A
Auxiliary material for Chapter 3

A.1. Derivation of the expression (3.78)

By applying the standard non-dimensionalization procedure mentioned in the relevant paragraph of Section 3.4.1 to the continuous system (2.18), the following expression is obtained:

$$\begin{cases} \partial_t \left(\rho \right) = \Psi_c^{(0)} \\ \partial_t \left(\rho u \right) = M_\star^{-2} \, \Theta_c^{(-2)} + \Theta_c^{(0)} \end{cases} \tag{A.1}$$

with:

$$\begin{cases} \Psi_c^{(0)} := - \partial_x \left(\rho u \right) \\ \Theta_c^{(-2)} := - \partial_x p \\ \Theta_c^{(0)} := - \partial_x \left(\rho u^2 \right) \end{cases} \tag{A.2}$$

and:

$$M_\star := \frac{u_{\mathrm{ref}}}{a_{\mathrm{ref}}} \tag{A.3}$$

The expressions (A.1) and (A.3) are copied in Section 3.4.1, respectively to (3.78) and (3.79), for ease of presentation.

A.2. Derivation of the expression (3.80)

By recalling the relevant definitions, the equation (3.72) can be recast as follows:

$$\begin{aligned}
2\,\mu_i \, \frac{\mathrm{d}}{\mathrm{d}t} \, \mathbf{q}_i^{(x)} \;=\;& \mathbf{f}_{i-1}^{(x)} - \mathbf{f}_{i+1}^{(x)} \\
& + \left| \tilde{\mathbf{J}}_{i(i+1)}^{(x)} \right| \cdot \Delta^{i(i+1)} \mathbf{q}^{(x)} \\
& - \left| \tilde{\mathbf{J}}_{(i-1)i}^{(x)} \right| \cdot \Delta^{(i-1)i} \mathbf{q}^{(x)} .
\end{aligned} \tag{A.4}$$

As a preliminary step, a suitable representation is sought for the generic term $\left| \tilde{\mathbf{J}}_{ij}^{(x)} \right| \cdot \Delta^{ij} \mathbf{q}^{(x)}$ appearing, in particular, in (A.4). To the purpose, once introduced the eigenvalue-eigenvector pairs of $\tilde{\mathbf{J}}_{ij}^{(x)}$, namely:

$$
\begin{cases}
\lambda_{ij}^{(1)} = u_{ij} + a_{ij} \quad , \quad \mathbf{r}_{ij}^{(1)} = \left(1, \lambda_{ij}^{(1)}\right)^T \\[3mm]
\lambda_{ij}^{(2)} = u_{ij} - a_{ij} \quad , \quad \mathbf{r}_{ij}^{(2)} = \left(1, \lambda_{ij}^{(2)}\right)^T
\end{cases}
\tag{A.5}
$$

it is possible to introduce the following equality:

$$
\left|\tilde{\mathbf{J}}_{ij}^{(x)}\right| \cdot \Delta^{ij} \mathbf{q}^{(x)} = \sum_{k=1}^{2} c_{ij}^{(k)} |\lambda_{ij}^{(k)}| \mathbf{r}_{ij}^{(k)}
\tag{A.6}
$$

where $c_{ij}^{(k)}$ denotes the k-th coordinate of $\Delta^{ij} \mathbf{q}^{(x)}$ with respect to the basis formed by the eigenvectors introduced in (A.5). Then, by exploiting the following classical property (see *e.g.* [1] or [111]):

$$
\Delta^{ij}(\rho u) = u_{ij} \, \Delta^{ij} \rho + \tilde{\rho}_{ij} \, \Delta^{ij} u
\tag{A.7}
$$

with:

$$
\tilde{\rho}_{ij} := \left(\rho_i \rho_j\right)^{1/2}
\tag{A.8}
$$

the following expressions are obtained:

$$
\begin{cases}
c_{ij}^{(1)} = \dfrac{1}{2a_{ij}} \left(\dfrac{\Delta^{ij} p}{a_{ij}} + \tilde{\rho}_{ij} \, \Delta^{ij} u\right) \\[5mm]
c_{ij}^{(2)} = \dfrac{1}{2a_{ij}} \left(\dfrac{\Delta^{ij} p}{a_{ij}} - \tilde{\rho}_{ij} \, \Delta^{ij} u\right)
\end{cases}
$$

and (A.6) can be recast as follows:

$$
\left|\tilde{\mathbf{J}}_{ij}^{(x)}\right| \cdot \Delta^{ij} \mathbf{q}^{(x)} = \dfrac{1}{2a_{ij}} \tilde{\mathbf{U}}_{ij}^{(x)} \cdot \begin{pmatrix} \Delta^{ij} p \\[3mm] \tilde{\rho}_{ij} \, \Delta^{ij} u \end{pmatrix}
\tag{A.9}
$$

where the components of the matrix $\tilde{\mathbf{U}}_{ij}^{(x)}$ read:

$$
\begin{cases}
\tilde{\mathbf{U}}_{ij}^{(x)}(1, 1) = \dfrac{|\lambda_{ij}^{(1)}| + |\lambda_{ij}^{(2)}|}{a_{ij}} \\[5mm]
\tilde{\mathbf{U}}_{ij}^{(x)}(1, 2) = |\lambda_{ij}^{(1)}| - |\lambda_{ij}^{(2)}| \\[5mm]
\tilde{\mathbf{U}}_{ij}^{(x)}(2, 1) = \dfrac{\lambda_{ij}^{(1)} |\lambda_{ij}^{(1)}| + \lambda_{ij}^{(2)} |\lambda_{ij}^{(2)}|}{a_{ij}} \\[5mm]
\tilde{\mathbf{U}}_{ij}^{(x)}(2, 2) = \lambda_{ij}^{(1)} |\lambda_{ij}^{(1)}| - \lambda_{ij}^{(2)} |\lambda_{ij}^{(2)}| .
\end{cases}
$$

For nearly-incompressible flows $|u_{ij}| \ll a_{ij}$ [1] and, consequently, the representation of $\tilde{U}_{ij}^{(x)}$ reduces to:

$$\tilde{U}_{ij}^{(x)} (M_\star \to 0) \to 2 \begin{pmatrix} 1 & u_{ij} \\ 2u_{ij} & u_{ij}^2 + a_{ij}^2 \end{pmatrix}. \qquad (A.10)$$

By substituting (A.10) into (A.9) and then back into the proper terms in (A.4), the following expression is obtained for the nearly-incompressible limit of the (dimensional) semi-discrete system at hand:

$$\begin{cases} 2\mu_i \dfrac{d}{dt} (\rho_i) &= \quad \Psi_{sd}^{(-1)} + \Psi_{sd}^{(0)} + \Psi_{sd}^{(1)} \\[3mm] 2\mu_i \dfrac{d}{dt} (\rho_i u_i) &= \Theta_{sd}^{(-2)} + \Theta_{sd}^{(-1)} + \Theta_{sd}^{(0)} + \Theta_{sd}^{(1)} \end{cases} \qquad (A.11)$$

where:

$$\begin{cases} \Psi_{sd}^{(-1)} &:= \dfrac{\Delta^{i(i+1)}p}{a_{i(i+1)}} - \dfrac{\Delta^{(i-1)i}p}{a_{(i-1)i}} \\[4mm] \Psi_{sd}^{(0)} &:= \rho_{i-1}u_{i-1} - \rho_{i+1}u_{i+1} \\[3mm] \Psi_{sd}^{(1)} &:= M_{i(i+1)} \, \tilde{\rho}_{i(i+1)} \, \Delta^{i(i+1)}u \\[3mm] & \quad - M_{(i-1)i} \, \tilde{\rho}_{(i-1)i} \, \Delta^{(i-1)i}u \\[3mm] \Theta_{sd}^{(-2)} &:= p_{i-1} - p_{i+1} \\[3mm] \Theta_{sd}^{(-1)} &:= 2 \left(M_{i(i+1)} \, \Delta^{i(i+1)}p - M_{(i-1)i} \, \Delta^{(i-1)i}p \right) \\[3mm] & \quad + a_{i(i+1)} \, \tilde{\rho}_{i(i+1)} \, \Delta^{i(i+1)}u - a_{(i-1)i} \, \tilde{\rho}_{(i-1)i} \, \Delta^{(i-1)i}u \\[3mm] \Theta_{sd}^{(0)} &:= \rho_{i-1}u_{i-1}^2 - \rho_{i+1}u_{i+1}^2 \\[3mm] \Theta_{sd}^{(1)} &:= M_{i(i+1)} \, u_{i(i+1)} \, \tilde{\rho}_{i(i+1)} \, \Delta^{i(i+1)}u \\[3mm] & \quad - M_{(i-1)i} \, u_{(i-1)i} \, \tilde{\rho}_{(i-1)i} \, \Delta^{(i-1)i}u \end{cases} \qquad (A.12)$$

---

[1] Indeed, the density is practically constant and therefore a_{ij}, as given by (3.62), is of the order of the characteristic sound speed $a_\star$ of the flow. On the other hand, due to the relevant convex combination in (3.61), $u_{ij} \leq \max(u_i, u_j) = a_\star O(M_\star)$.

and:

$$M_{ij} := \frac{u_{ij}}{a_{ij}}.$$ (A.13)

Finally, by applying the standard non-dimensionalization procedure mentioned in the relevant paragraph of Section 3.4.1 to the system (A.11), the following expression is obtained:

$$
\begin{cases}
2\mu_i \dfrac{d}{dt}(\rho_i) &= & M_\star^{-1}\Psi_{sd}^{(-1)} \;+\; \hat{\Psi}_{sd}^{(0)} \\[2mm]
2\mu_i \dfrac{d}{dt}(\rho_i u_i) &= M_\star^{-2}\Theta_{sd}^{(-2)} \;+\; M_\star^{-1}\Theta_{sd}^{(-1)} \;+\; \hat{\Theta}_{sd}^{(0)}
\end{cases}
$$ (A.14)

where:

$$
\begin{cases}
\hat{\Psi}_{sd}^{(0)} &:= \; \Psi_{sd}^{(0)} + M_\star \Psi_{sd}^{(1)} \\[2mm]
\hat{\Theta}_{sd}^{(0)} &:= \; \Theta_{sd}^{(0)} + M_\star \Theta_{sd}^{(1)}
\end{cases}
$$ (A.15)

and the relevant coefficients are recalled from (A.12). The system (A.14) is copied in Section 3.4.1, namely to (3.80), for ease of presentation.

A.3. Proof of the Proposition 3.4.4 (Section 3.4.1)

By substituting the expansion of the continuous solution (3.81) into the relevant system (3.78), the following relations are obtained:

$$
\begin{cases}
\partial_t \rho_0 &= & \dddot{\Psi}_c^{(0)} \\[2mm]
\partial_t (\rho_0 u_0) &= M_\star^{-2}\check{\Theta}_c^{(-2)} \;+\; M_\star^{-1}\check{\Theta}_c^{(-1)} \;+\; \ddot{\Theta}_c^{(0)}
\end{cases}
$$ (A.16)

where:

$$
\begin{cases}
\dddot{\Psi}_c^{(0)} &:= \; -\partial_x (\rho_0 u_0) + M_\star (\cdots) \\[2mm]
\check{\Theta}_c^{(-2)} &:= \; -\partial_x p_0 \\[2mm]
\check{\Theta}_c^{(-1)} &:= \; -\partial_x p_1 \\[2mm]
\ddot{\Theta}_c^{(0)} &:= \; -\partial_x \left(\rho_0 u_0^2 + p_2\right) + M_\star (\cdots).
\end{cases}
$$ (A.17)

Clearly, it is possible to solve the system (A.16) for $M_\star \to 0$ only if:

$$\check{\Theta}_c^{(-2)} = 0 \quad, \quad \check{\Theta}_c^{(-1)} = 0.$$ (A.18)

The equations (A.18) imply that:

$$p_0(x, t) = \bar{p}_0(t) \quad , \quad p_1(x, t) = \bar{p}_1(t)$$

for suitable functions $\bar{p}_0$ and $\bar{p}_1$ and therefore the asymptotic expression (3.84) is obtained.

As far as the semi-discrete problem is concerned, by substituting the expansion (3.82) into the relevant system (3.80), the following relations are obtained:

$$\begin{cases} 2\mu_i \dfrac{d}{dt}(\rho_{0i}) & = & M_\star^{-1}\,\check{\Psi}_{sd}^{(-1)} + \ddot{\Psi}_{sd}^{(0)} \\[4mm] 2\mu_i \dfrac{d}{dt}(\rho_{0i}u_{0i}) & = & M_\star^{-2}\,\check{\Theta}_{sd}^{(-2)} + M_\star^{-1}\,\check{\Theta}_{sd}^{(-1)} + \ddot{\Theta}_{sd}^{(0)} \end{cases} \quad (A.19)$$

where the coefficients are suitably defined in terms of the entities introduced in (A.12) (the definitions are not reported here because inessential to the present purposes). As for the continuous case, the coefficients associated with the negative powers of $M_\star$ in (A.19) must be identically equal to zero in order to allow for the solution to be defined when $M_\star \to 0$. In particular, the following relations are obtained by respectively imposing $\check{\Psi}_{sd}^{(-1)} = 0$ and $\check{\Theta}_{sd}^{(-2)} = 0$:

$$\frac{\Delta^{i(i+1)}p_0}{a_{0i(i+1)}} - \frac{\Delta^{(i-1)i}p_0}{a_{0(i-1)i}} = 0 \quad (A.20)$$

$$\Delta^{i(i+1)}p_0 + \Delta^{(i-1)i}p_0 = 0. \quad (A.21)$$

The equations (A.20) and (A.21) above only admit the following solution (as usual, $i \in \mathcal{I}$ and $j \in \pi_i$):

$$\Delta^{ij}p_0 = 0. \quad (A.22)$$

Indeed, according to (A.20), $\Delta^{i(i+1)}p_0$ has the same sign as $\Delta^{(i-1)i}p_0$ (a_{0ij} being positive), in contrast with (A.21) unless (A.22) holds. The relation (A.22), in turn, implies that:

$$p_{0i}(t) = \tilde{p}_0(t)$$

for a certain function $\tilde{p}_0$. In consideration of (A.22), the condition $\check{\Theta}_{sd}^{(-1)} = 0$ leads to the following relation:

$$\tilde{\rho}_0\left(a_{0i(i+1)}\Delta^{i(i+1)}u_0 - a_{0(i-1)i}\Delta^{(i-1)i}u_0\right) + p_{1(i+1)} - p_{1(i-1)} = 0 \quad (A.23)$$

where $\tilde{\rho}_0$ is defined by inverting the first relation in (3.83) in correspondence of $p_0 = \tilde{p}_0$. The equation (A.23), in general, does not impose any specific constraint on $p_{1i}(t)$. As a result, the asymptotic expression (3.85) is recovered. This completes the proof. $\qquad\square$

A.4. Derivation of the expression (3.96)

By recalling the relevant definitions, the equation (3.95) can be recast as follows:

$$
2\mu_i \frac{d}{dt} \mathbf{q}_i^{(x)} = \mathbf{f}_{i-1}^{(x)} - \mathbf{f}_{i+1}^{(x)}
$$

$$
+ \left(\mathbf{P}_{i(i+1)}^{(x)}\right)^{-1} \cdot \left| \mathbf{P}_{i(i+1)}^{(x)} \cdot \tilde{\mathbf{J}}_{i(i+1)}^{(x)} \right| \cdot \Delta^{i(i+1)} \mathbf{q}^{(x)} \qquad \text{(A.24)}
$$

$$
- \left(\mathbf{P}_{(i-1)i}^{(x)}\right)^{-1} \cdot \left| \mathbf{P}_{(i-1)i}^{(x)} \cdot \tilde{\mathbf{J}}_{(i-1)i}^{(x)} \right| \cdot \Delta^{(i-1)i} \mathbf{q}^{(x)} .
$$

As a preliminary step, a suitable representation is sought for the generic term $(\mathbf{P}_{ij}^{(x)})^{-1} \cdot |\mathbf{P}_{ij}^{(x)} \cdot \tilde{\mathbf{J}}_{ij}^{(x)}| \cdot \Delta^{ij} \mathbf{q}^{(x)}$ appearing, in particular, in (A.24). To the purpose, once introduced the eigenvalue-eigenvector pairs of $(\mathbf{P}_{ij}^{(x)} \cdot \tilde{\mathbf{J}}_{ij}^{(x)})$, namely:

$$
\begin{cases}
\lambda_{ij}^{(1,p)} = u_{ij}^{(p)} + a_{ij}^{(p)} & , \quad \mathbf{r}_{ij}^{(1,p)} = \left(1, \lambda_{ij}^{(1,p)}\right)^T \\[2ex]
\lambda_{ij}^{(2,p)} = u_{ij}^{(p)} - a_{ij}^{(p)} & , \quad \mathbf{r}_{ij}^{(2,p)} = \left(1, \lambda_{ij}^{(2,p)}\right)^T
\end{cases}
\qquad \text{(A.25)}
$$

with:

$$
u_{ij}^{(p)} := \frac{1+\beta^2}{2} u_{ij}
$$

and:

$$
a_{ij}^{(p)} := \left(\left(\frac{1-\beta^2}{2} u_{ij}\right)^2 + (\beta\, a_{ij})^2 \right)^{\frac{1}{2}} \qquad \text{(A.26)}
$$

it is possible to introduce the following equality:

$$
\left(\mathbf{P}_{ij}^{(x)}\right)^{-1} \cdot \left| \mathbf{P}_{ij}^{(x)} \cdot \tilde{\mathbf{J}}_{ij}^{(x)} \right| \cdot \Delta^{ij} \mathbf{q}^{(x)} = \left(\mathbf{P}_{ij}^{(x)}\right)^{-1} \cdot \left(\sum_{k=1}^{2} c_{ij}^{(k,p)} \, |\lambda_{ij}^{(k,p)}| \, \mathbf{r}_{ij}^{(k,p)} \right)
$$

$$\text{(A.27)}$$

where $c_{ij}^{(k,p)}$ denotes the k-th coordinate of $\Delta^{ij} \mathbf{q}^{(x)}$ with respect to the basis formed by the eigenvectors introduced in (A.25). Then, by exploiting the classical property (A.7), the following expressions are obtained:

$$
\begin{cases}
c_{ij}^{(1,p)} = \dfrac{\beta^2}{2\, a_{ij}^{(p)}} \left(\dfrac{\Delta^{ij} p}{\sigma_{ij}} + \tilde{\rho}_{ij}\, \Delta^{ij} u \right) \\[3ex]
c_{ij}^{(2,p)} = \dfrac{\beta^2}{2\, a_{ij}^{(p)}} \left(\dfrac{\Delta^{ij} p}{\tau_{ij}} - \tilde{\rho}_{ij}\, \Delta^{ij} u \right)
\end{cases}
$$

with:

$$\sigma_{ij} := a_{ij}^{(p)} + \frac{1-\beta^2}{2} u_{ij} \quad , \quad \tau_{ij} := a_{ij}^{(p)} - \frac{1-\beta^2}{2} u_{ij}$$

and (A.27) can be recast as follows:

$$\left(\mathbf{P}_{ij}^{(x)}\right)^{-1} \cdot \left|\mathbf{P}_{ij}^{(x)} \cdot \tilde{\mathbf{J}}_{ij}^{(x)}\right| \cdot \Delta^{ij} \mathbf{q}^{(x)} = \frac{1}{2 a_{ij}^{(p)}} \, \tilde{\mathbf{U}}_{ij}^{(x)p} \cdot \begin{pmatrix} \Delta^{ij} p \\ \tilde{\rho}_{ij} \Delta^{ij} u \end{pmatrix} \quad \text{(A.28)}$$

where the components of the matrix $\tilde{\mathbf{U}}_{ij}^{(x)p}$ read:

$$\begin{cases} \tilde{U}_{ij}^{(x)p}(1,1) = \dfrac{|\lambda_{ij}^{(1,p)}|}{\sigma_{ij}} + \dfrac{|\lambda_{ij}^{(2,p)}|}{\tau_{ij}} \\[2ex] \tilde{U}_{ij}^{(x)p}(1,2) = |\lambda_{ij}^{(1,p)}| - |\lambda_{ij}^{(2,p)}| \\[2ex] \tilde{U}_{ij}^{(x)p}(2,1) = \dfrac{\lambda_{ij}^{(1,p)}|\lambda_{ij}^{(1,p)}|}{\sigma_{ij}} + \dfrac{\lambda_{ij}^{(2,p)}|\lambda_{ij}^{(2,p)}|}{\tau_{ij}} + (1-\beta^2)u_{ij}\tilde{U}_{ij}^{(x)p}(1,1) \\[2ex] \tilde{U}_{ij}^{(x)p}(2,2) = \lambda_{ij}^{(1,p)}|\lambda_{ij}^{(1,p)}| - \lambda_{ij}^{(2,p)}|\lambda_{ij}^{(2,p)}| + \left(1-\beta^2\right)u_{ij}\tilde{U}_{ij}^{(x)p}(1,2). \end{cases}$$

For nearly-incompressible flows $|u_{ij}^{(p)}| \ll a_{ij}^{(p)}$; indeed:

$$\left(u_{ij}^{(p)}\right)^2 - \left(a_{ij}^{(p)}\right)^2 = \beta^2 \left(u_{ij}^2 - a_{ij}^2\right)$$

and therefore the considerations already introduced when discussing the non-preconditioned case in Section A.2 can be applied. Consequently, the representation of the matrix $\tilde{\mathbf{U}}_{ij}^{(x)p}$ in (A.28) reduces to:

$$\tilde{\mathbf{U}}_{ij}^{(x)p} \, (M_\star \to 0) \to 2 \begin{pmatrix} 1 - \dfrac{1-\beta^2}{2}M_{ij}^2 & \dfrac{1+\beta^2}{2}u_{ij} \\[2ex] \left(\dfrac{3+\beta^2}{2} - \dfrac{1-\beta^2}{2}M_{ij}^2\right)u_{ij} & u_{ij}^2 + \beta^2 a_{ij}^2 \end{pmatrix}$$

$$\text{(A.29)}$$

where M_{ij} is defined in (A.13). By substituting (A.29) into (A.28) and then back into the proper terms in (A.24), the following expression is obtained for the nearly-incompressible limit of the (dimensional) semi-discrete system at hand:

$$
\begin{cases}
2\mu_i \dfrac{d}{dt}(\rho_i) = \Psi^{(-1)}_{sd,p} + \Psi^{(0)}_{sd,p} + \Psi^{(1)}_{sd,p} \\[2mm]
2\mu_i \dfrac{d}{dt}(\rho_i u_i) = \Theta^{(-2)}_{sd,p} + \Theta^{(-1)}_{sd,p} + \Theta^{(0)}_{sd,p} + \Theta^{(1)}_{sd,p}
\end{cases}
\tag{A.30}
$$

where:

$$
\begin{cases}
\Psi^{(-1)}_{sd,p} := \dfrac{\Delta^{i(i+1)}p}{a^{(p)}_{i(i+1)}} - \dfrac{\Delta^{(i-1)i}p}{a^{(p)}_{(i-1)i}} \\[3mm]
\Psi^{(0)}_{sd,p} := \rho_{i-1}u_{i-1} - \rho_{i+1}u_{i+1} \\[3mm]
\Psi^{(1)}_{sd,p} := \dfrac{1+\beta^2}{2}\left(M^{(p)}_{i(i+1)}\tilde{\rho}_{i(i+1)}\Delta^{i(i+1)}u - M^{(p)}_{(i-1)i}\tilde{\rho}_{(i-1)i}\Delta^{(i-1)i}u \right) \\[3mm]
\qquad -\dfrac{1-\beta^2}{2}\left(\left(M_{i(i+1)}\right)^2 \dfrac{\Delta^{i(i+1)}p}{a^{(p)}_{i(i+1)}} - \left(M_{(i-1)i}\right)^2 \dfrac{\Delta^{(i-1)i}p}{a^{(p)}_{(i-1)i}} \right) \\[3mm]
\Theta^{(-2)}_{sd,p} := p_{i-1} - p_{i+1} \\[3mm]
\Theta^{(-1)}_{sd,p} := \dfrac{3+\beta^2}{2}\left(M^{(p)}_{i(i+1)}\Delta^{i(i+1)}p - M^{(p)}_{(i-1)i}\Delta^{(i-1)i}p \right) \\[3mm]
\qquad +\beta^2 \left(\dfrac{a^2_{i(i+1)}}{a^{(p)}_{i(i+1)}}\tilde{\rho}_{i(i+1)}\Delta^{i(i+1)}u - \dfrac{a^2_{(i-1)i}}{a^{(p)}_{(i-1)i}}\tilde{\rho}_{(i-1)i}\Delta^{(i-1)i}u \right) \\[3mm]
\Theta^{(0)}_{sd,p} := \rho_{i-1}u^2_{i-1} - \rho_{i+1}u^2_{i+1} \\[3mm]
\Theta^{(1)}_{sd,p} := M^{(p)}_{i(i+1)}u_{i(i+1)}\tilde{\rho}_{i(i+1)}\Delta^{i(i+1)}u - M^{(p)}_{(i-1)i}u_{(i-1)i}\tilde{\rho}_{(i-1)i}\Delta^{(i-1)i}u \\[3mm]
\qquad -\dfrac{1-\beta^2}{2}\left(\left(M_{i(i+1)}\right)^2 M^{(p)}_{i(i+1)}\Delta^{i(i+1)}p \right. \\[3mm]
\qquad \left. - \left(M_{(i-1)i}\right)^2 M^{(p)}_{(i-1)i}\Delta^{(i-1)i}p \right)
\end{cases}
\tag{A.31}
$$

and:

$$
M^{(p)}_{ij} := \dfrac{u_{ij}}{a^{(p)}_{ij}}.
$$

The non-dimensional counterpart of the system (A.30) is obtained by applying the standard non-dimensionalization procedure mentioned in the relevant paragraph of Section 3.4.1. It is worth remarking that, to the purpose, the reference sound speed a_{ref} introduced in (3.77) is exploited for non-dimensionalizing $a^{(p)}_{ij}$, in view of the fact that $a^{(p)}_{ij}\left(\beta^2 \to 1\right) \to a_{ij}$.

Hence, the non-dimensional counterpart of (A.26) in particular reads:[2]

$$a_{ij}^{(p)} = \left(\left(\frac{1 - \beta^2}{2} M_\star u_{ij} \right)^2 + \left(\beta a_{ij} \right)^2 \right)^{\frac{1}{2}} \qquad \text{(A.32)}$$

Then, by assuming that the parameter β is formally of the order of the unity, the following non-dimensional expression is obtained for the system (A.30):

$$\begin{cases} 2\mu_i \dfrac{d}{dt} (\rho_i) & = & M_\star^{-1} \Psi_{sd,p}^{(-1)} + \hat{\Psi}_{sd,p}^{(0)} \\[4mm] 2\mu_i \dfrac{d}{dt} (\rho_i u_i) & = & M_\star^{-2} \Theta_{sd,p}^{(-2)} + M_\star^{-1} \Theta_{sd,p}^{(-1)} + \hat{\Theta}_{sd,p}^{(0)} \end{cases} \qquad \text{(A.33)}$$

where:

$$\begin{cases} \hat{\Psi}_{sd,p}^{(0)} & := & \Psi_{sd,p}^{(0)} + M_\star \Psi_{sd,p}^{(1)} \\[4mm] \hat{\Theta}_{sd,p}^{(0)} & := & \Theta_{sd,p}^{(0)} + M_\star \Theta_{sd,p}^{(1)} \end{cases} \qquad \text{(A.34)}$$

and the relevant coefficients (of course, here understood as non-dimensional) are recalled from (A.31). The system (A.33) is copied in Section 3.4.1, namely to (3.96), for ease of presentation.

A.5. A remark on the expression (3.98)

Both the (non-dimensional) systems (3.96) and (3.98) derive from the (dimensional) system (A.30) by the non-dimensionalization procedure mentioned in the relevant paragraph of Section 3.4.1. More in detail, in the former case β is assumed of the order of the unity while in the latter one it is assumed of the order of $M_\star$ by means of the position (3.97). Clearly, (3.97) directly affects the considered non-dimensional equations through the definitions of the relevant coefficients associated with the powers of $M_\star$. As an example, the non-dimensional form of $a_{ij}^{(p)}$, given by (A.32), can be considered. Once introduced the following definition:

$$\alpha_{ij} := \left(a_{ij}^{(p)} \right)^2$$

the expansion of $\left(a_{ij}^{(p)} \right)^{-1}$, which appears in (A.31) both directly and via $M_{ij}^{(p)}$, varies as shown below:

---

[2] The same symbol is used for corresponding dimensional and non-dimensional entities, as declared in Note 3.4.1 (Section 3.4.1).

- without the position (3.97) the expansion of α_{ij} reads:

$$\alpha_{ij} = \alpha_{0ij} + M_\star \, \alpha_{1ij} + \cdots$$

where:

$$\begin{cases} \alpha_{0ij} &= \beta^2 \, a_{0ij}^2 \\[2mm] \alpha_{1ij} &= 2\beta^2 \, a_{0ij} \, a_{1ij} \end{cases}$$

and therefore (α_{0ij} is positive, see Note 3.4.3 in Section 3.4.1):

$$\left(a_{ij}^{(p)}\right)^{-1} = \alpha_{0ij}^{-\frac{1}{2}} \left(1 - M_\star \, \frac{\alpha_{1ij}}{2\,\alpha_{0ij}} + \cdots\right)$$

- with the position (3.97), the following expansion must be considered:

$$\alpha_{ij} = M_\star^2 \left(\alpha_{2ij} + M_\star \, \alpha_{3ij} + \cdots\right)$$

where:

$$\begin{cases} \alpha_{2ij} &= \beta_{\text{ref}}^2 \, a_{0ij}^2 + \dfrac{1}{4} u_{0ij}^2 \\[3mm] \alpha_{3ij} &= 2\,\beta_{\text{ref}}^2 \, a_{0ij} \, a_{1ij} + \dfrac{1}{2} u_{0ij} \, u_{1ij} \,. \end{cases} \tag{A.35}$$

and thus (α_{2ij} is clearly positive, see Note 3.4.3 in Section 3.4.1):

$$\left(a_{ij}^{(p)}\right)^{-1} = M_\star^{-1} \, \alpha_{2ij}^{-\frac{1}{2}} \left(1 - M_\star \, \frac{\alpha_{3ij}}{2\,\alpha_{2ij}} + \cdots\right)$$

A.6. Proof of the Proposition 3.4.9 (Section 3.4.3)

The coefficients associated with the negative powers of $M_\star$ in (3.98) must be identically equal to zero in order to allow for the solution to be defined for $M_\star \to 0$. In particular, the following relations are derived by respectively imposing $\check{\Psi}_{sd,p}^{(-2)} = 0$ and $\check{\Theta}_{sd,p}^{(-2)} = 0$:

$$\frac{\Delta^{i(i+1)} p_0}{\alpha_{2i(i+1)}^{1/2}} - \frac{\Delta^{(i-1)i} p_0}{\alpha_{2(i-1)i}^{1/2}} = 0 \tag{A.36}$$

$$\Delta^{i(i+1)} p_0 - \Delta^{(i-1)i} p_0$$

$$+ \frac{3}{2}\left(\frac{u_{0i(i+1)}}{\alpha_{2i(i+1)}^{1/2}} \Delta^{i(i+1)} p_0 - \frac{u_{0(i-1)i}}{\alpha_{2(i-1)i}^{1/2}} \Delta^{(i-1)i} p_0\right) = 0. \tag{A.37}$$

with α_{2ij} given by (A.35). The equations (A.36) and (A.37) constitute a system of two homogeneous difference equations for the two unknowns $\Delta^{(i-1)i} p_0$ and $\Delta^{i(i+1)} p_0$. Due to the arbitrariness of the coefficients (depending on the solution itself), it necessarily follows that (as usual, $i \in \mathcal{I}$ and $j \in \pi_i$):

$$\Delta^{ij} p_0 = 0 \qquad (A.38)$$

and therefore p_{0i} does not depend on the spatial index i:

$$p_{0i}(t) = \hat{p}_0(t) \qquad (A.39)$$

for a suitable function $\hat{p}_0$. In consideration of (A.38), the conditions $\check{\Psi}_{sd,p}^{(-1)} = 0$ and $\check{\Theta}_{sd,p}^{(-1)} = 0$ respectively read:

$$\frac{\Delta^{i(i+1)} p_1}{\alpha_{2i(i+1)}^{1/2}} - \frac{\Delta^{(i-1)i} p_1}{\alpha_{2(i-1)i}^{1/2}} = 0 \qquad (A.40)$$

$$\Delta^{i(i+1)} p_1 - \Delta^{(i-1)i} p_1$$

$$+ \frac{3}{2} \left(\frac{u_{0i(i+1)}}{\alpha_{2i(i+1)}^{1/2}} \Delta^{i(i+1)} p_1 - \frac{u_{0(i-1)i}}{\alpha_{2(i-1)i}^{1/2}} \Delta^{(i-1)i} p_1 \right) = 0. \qquad (A.41)$$

The equations (A.40) and (A.41) are identical to (A.36) and (A.37), respectively, once replaced p_1 with p_0. Hence, the following condition can be immediately drawn from (A.39):

$$p_{1i}(t) = \hat{p}_1(t)$$

where $\hat{p}_1$ denotes a suitable function. As a result, the asymptotic expression (3.99) is recovered. This completes the proof. □

A.7. Proof of the Proposition 3.5.2 (Section 3.5.1)

In order to simplify the notation, a generic function $\mathbf{g}(\mathbf{u}, \mathbf{v})$ is considered at a preliminary stage, $\mathbf{u}$ and $\mathbf{v}$ hereafter denoting generic vectors. Moreover, the following definitions, directly derived from (3.128), are introduced:

$$\begin{cases} \Delta_{(L)}\mathbf{g} := \mathbf{g}(\mathbf{u}, \mathbf{v}_0) - \mathbf{g}(\mathbf{u}_0, \mathbf{v}_0) \\[2mm] \Delta_{(R)}\mathbf{g} := \mathbf{g}(\mathbf{u}_0, \mathbf{v}) - \mathbf{g}(\mathbf{u}_0, \mathbf{v}_0) \\[2mm] \bar{\Delta}_{(L)}\mathbf{g} := \mathbf{g}(\mathbf{u}_0, \mathbf{v}) - \mathbf{g}(\mathbf{u}, \mathbf{v}) \\[2mm] \bar{\Delta}_{(R)}\mathbf{g} := \mathbf{g}(\mathbf{u}, \mathbf{v}_0) - \mathbf{g}(\mathbf{u}, \mathbf{v}) \end{cases}$$

where $\mathbf{u}_0$ and $\mathbf{v}_0$ represent specific instances of $\mathbf{u}$ and $\mathbf{v}$, respectively.

The proof under consideration exploits the algebraic relation described by the subsequent:

Lemma A.7.1. *Let* $\mathbf{M}(\cdot, \cdot)$ *and* $\mathbf{N}(\cdot, \cdot)$ *denote suitable matrices and let* $\hat{\mathbf{r}}(\cdot, \cdot, \cdot, \cdot)$ *represent a suitable vector, such that:*

$$\Delta_{(L)}\mathbf{g} + \Delta_{(R)}\mathbf{g} = \mathbf{M}(\mathbf{u}, \mathbf{v}_0) \cdot (\mathbf{u} - \mathbf{u}_0)$$

$$+ \mathbf{N}(\mathbf{u}_0, \mathbf{v}) \cdot (\mathbf{v} - \mathbf{v}_0) \tag{A.42}$$

$$+ \hat{\mathbf{r}}(\mathbf{u}_0, \mathbf{v}_0, \mathbf{u}, \mathbf{v})$$

for any value of $\mathbf{u}_0$, $\mathbf{v}_0$, $\mathbf{u}$ *and* $\mathbf{v}$. *Then, the following relation is satisfied:*

$$\mathbf{g}(\mathbf{u}, \mathbf{v}) - \mathbf{g}(\mathbf{u}_0, \mathbf{v}_0) = \mathbf{M}(\mathbf{u}_0, \mathbf{v}_0) \cdot (\mathbf{u} - \mathbf{u}_0)$$

$$+ \mathbf{N}(\mathbf{u}_0, \mathbf{v}_0) \cdot (\mathbf{v} - \mathbf{v}_0) \tag{A.43}$$

$$+ \frac{1}{2}\mathbf{r}(\mathbf{u}_0, \mathbf{v}_0, \mathbf{u}, \mathbf{v})$$

with:

$$\mathbf{r}(\mathbf{u}_0, \mathbf{v}_0, \mathbf{u}, \mathbf{v}) := \left(\Delta_{(L)}\mathbf{M} + \Delta_{(R)}\mathbf{M}\right) \cdot (\mathbf{u} - \mathbf{u}_0)$$

$$+ \left(\Delta_{(L)}\mathbf{N} + \Delta_{(R)}\mathbf{N}\right) \cdot (\mathbf{v} - \mathbf{v}_0)$$

$$- \Delta_{(LR)}\hat{\mathbf{r}}(\cdot, \cdot, \mathbf{u}_0, \mathbf{v})$$

$$- \Delta_{(LR)}\hat{\mathbf{r}}(\cdot, \cdot, \mathbf{u}, \mathbf{v}_0)$$

and:

$$\Delta_{(LR)}\hat{\mathbf{r}}(\cdot, \cdot, \bar{\mathbf{u}}, \bar{\mathbf{v}}) := \hat{\mathbf{r}}(\mathbf{u}, \mathbf{v}, \bar{\mathbf{u}}, \bar{\mathbf{v}}) - \hat{\mathbf{r}}(\mathbf{u}_0, \mathbf{v}_0, \bar{\mathbf{u}}, \bar{\mathbf{v}}) \ .$$

Proof. By firstly choosing $\mathbf{v} = \mathbf{v}_0$ and then $\mathbf{u} = \mathbf{u}_0$ in (A.42), the following expressions are respectively obtained:

$$\Delta_{(L)}\mathbf{g} = \mathbf{M}(\mathbf{u}, \mathbf{v}_0) \cdot (\mathbf{u} - \mathbf{u}_0) + \hat{\mathbf{r}}(\mathbf{u}_0, \mathbf{v}_0, \mathbf{u}, \mathbf{v}_0) \tag{A.44}$$

$$\Delta_{(R)}\mathbf{g} = \mathbf{N}(\mathbf{u}_0, \mathbf{v}) \cdot (\mathbf{v} - \mathbf{v}_0) + \hat{\mathbf{r}}(\mathbf{u}_0, \mathbf{v}_0, \mathbf{u}_0, \mathbf{v}) \ . \tag{A.45}$$

Furthermore, by inverting the role of $(\mathbf{u}_0, \mathbf{v}_0)$ and $(\mathbf{u}, \mathbf{v})$ it follows that:

$$\bar{\Delta}_{(L)}\mathbf{g} = \mathbf{M}(\mathbf{u}_0, \mathbf{v}) \cdot (\mathbf{u}_0 - \mathbf{u}) + \hat{\mathbf{r}}(\mathbf{u}, \mathbf{v}, \mathbf{u}_0, \mathbf{v}) \tag{A.46}$$

$$\bar{\Delta}_{(R)}\mathbf{g} = \mathbf{N}(\mathbf{u}, \mathbf{v}_0) \cdot (\mathbf{v}_0 - \mathbf{v}) + \hat{\mathbf{r}}(\mathbf{u}, \mathbf{v}, \mathbf{u}, \mathbf{v}_0) \ . \tag{A.47}$$

Then, once noticed that:

$$2\left(\mathbf{g}\left(\mathbf{u},\mathbf{v}\right)-\mathbf{g}\left(\mathbf{u}_0,\mathbf{v}_0\right)\right)=\Delta_{(L)}\mathbf{g}+\Delta_{(R)}\mathbf{g}-\bar{\Delta}_{(L)}\mathbf{g}-\bar{\Delta}_{(R)}\mathbf{g}\quad\text{(A.48)}$$

the equality (A.43) is immediately obtained by substituting (A.44)-(A.47) into the corresponding terms on the right-hand side of (A.48). This completes the proof. □

Let $\boldsymbol{\phi}^{\mathrm{ROE}}\left(\mathbf{u},\mathbf{v}\right)$ denote, in the present context, the generic Roe numerical flux $\boldsymbol{\phi}_{LR}^{(g)ROE}$ considered in Section 3.5.1. It is possible to recast the considered numerical flux as follows (from (3.43), by a trivial change of notation):

$$\begin{cases} \boldsymbol{\phi}^{\mathrm{ROE}}\left(\mathbf{u},\mathbf{v}\right)=\mathbf{f}\left(\mathbf{u}\right)+\tilde{\mathbf{J}}^-\left(\mathbf{u},\mathbf{v}\right)\cdot\left(\mathbf{v}-\mathbf{u}\right) \\[2mm] \boldsymbol{\phi}^{\mathrm{ROE}}\left(\mathbf{u},\mathbf{v}\right)=\mathbf{f}\left(\mathbf{v}\right)-\tilde{\mathbf{J}}^+\left(\mathbf{u},\mathbf{v}\right)\cdot\left(\mathbf{v}-\mathbf{u}\right). \end{cases}\quad\text{(A.49)}$$

Then, from the first relation in (A.49) it follows that:

$$\Delta_{(R)}\boldsymbol{\phi}^{\mathrm{ROE}}=\tilde{\mathbf{J}}^-\left(\mathbf{u}_0,\mathbf{v}\right)\cdot\left(\mathbf{v}-\mathbf{u}_0\right)-\tilde{\mathbf{J}}^-\left(\mathbf{u}_0,\mathbf{v}_0\right)\cdot\left(\mathbf{v}_0-\mathbf{u}_0\right)\quad\text{(A.50)}$$

while from the second one it follows that:

$$\Delta_{(L)}\boldsymbol{\phi}^{\mathrm{ROE}}=\tilde{\mathbf{J}}^+\left(\mathbf{u}_0,\mathbf{v}_0\right)\cdot\left(\mathbf{v}_0-\mathbf{u}_0\right)-\tilde{\mathbf{J}}^+\left(\mathbf{u},\mathbf{v}_0\right)\cdot\left(\mathbf{v}_0-\mathbf{u}\right).\quad\text{(A.51)}$$

Moreover, by combining (A.50) and (A.51) the subsequent relation is obtained:

$$\Delta_{(L)}\boldsymbol{\phi}^{\mathrm{ROE}}+\Delta_{(R)}\boldsymbol{\phi}^{\mathrm{ROE}}=\tilde{\mathbf{J}}^+\left(\mathbf{u},\mathbf{v}_0\right)\cdot\left(\mathbf{u}-\mathbf{u}_0\right)$$

$$+\tilde{\mathbf{J}}^-\left(\mathbf{u}_0,\mathbf{v}\right)\cdot\left(\mathbf{v}-\mathbf{v}_0\right)\quad\text{(A.52)}$$

$$+\hat{\mathbf{r}}^{\mathrm{ROE}}\left(\mathbf{u}_0,\mathbf{v}_0,\mathbf{u},\mathbf{v}\right)$$

with:

$$\hat{\mathbf{r}}^{\mathrm{ROE}}\left(\mathbf{u}_0,\mathbf{v}_0,\mathbf{u},\mathbf{v}\right):=\left(\Delta_{(R)}\tilde{\mathbf{J}}^--\Delta_{(L)}\tilde{\mathbf{J}}^+\right)\cdot\left(\mathbf{v}_0-\mathbf{u}_0\right).$$

In consideration of the similarity between (A.42) and (A.52), it is possible to apply the Lemma A.7.1 introduced above, thus obtaining the following relation:

$$\boldsymbol{\phi}^{\mathrm{ROE}}\left(\mathbf{u},\mathbf{v}\right)-\boldsymbol{\phi}^{\mathrm{ROE}}\left(\mathbf{u}_0,\mathbf{v}_0\right)=\tilde{\mathbf{J}}^+\left(\mathbf{u}_0,\mathbf{v}_0\right)\cdot\left(\mathbf{u}-\mathbf{u}_0\right)$$

$$+\tilde{\mathbf{J}}^-\left(\mathbf{u}_0,\mathbf{v}_0\right)\cdot\left(\mathbf{v}-\mathbf{v}_0\right)\quad\text{(A.53)}$$

$$+\frac{1}{2}\mathbf{r}^{\mathrm{ROE}}\left(\mathbf{u}_0,\mathbf{v}_0,\mathbf{u},\mathbf{v}\right)$$

with:

$$\mathbf{r}^{\text{ROE}}(\mathbf{u}_0, \mathbf{v}_0, \mathbf{u}, \mathbf{v}) := \left(\Delta_{(L)}\tilde{\mathbf{J}}^+ + \Delta_{(R)}\tilde{\mathbf{J}}^+\right) \cdot (\mathbf{u} - \mathbf{u}_0)$$

$$+ \left(\Delta_{(L)}\tilde{\mathbf{J}}^- + \Delta_{(R)}\tilde{\mathbf{J}}^-\right) \cdot (\mathbf{v} - \mathbf{v}_0)$$

$$+ \left(\Delta_{(R)}\tilde{\mathbf{J}}^- - \Delta_{(L)}\tilde{\mathbf{J}}^+\right) \cdot (\mathbf{v}_0 - \mathbf{u}_0)$$

$$+ \left(\bar{\Delta}_{(L)}\tilde{\mathbf{J}}^+ - \bar{\Delta}_{(R)}\tilde{\mathbf{J}}^-\right) \cdot (\mathbf{v} - \mathbf{u})$$

Finally, the equality (3.126) is directly obtained from (A.53) by means of a straightforward change of notation. This completes the proof. □

Appendix B
Efficient access to the table (4.10) for pressure-based algorithms

A typical access to the table (4.10) within a pressure-based algorithm is aimed at finding the density ρ and the sound speed a corresponding to a certain input value of the independent variable $p < p_{\text{sat}}$.

It is possible to define a fast look-up strategy by firstly noticing that the distribution along the y-axis in Figure 4.1 of the pressure "nodes" p_i, as provided by an ordinary adaptive integration algorithm (*e.g.* a classical fourth-order Runge-Kutta scheme [79] with adaptive step-size control), is typically clustered around a node p_{i_*} corresponding to a density ρ_{i_*} such that $\rho_{i_*} \approx 0.5 \cdot \rho_{\text{Lsat}}$. It is therefore possible to approximate the original pressure sequence p_i by a new one, say p_i'', obtained by juxtaposing two geometric sequences, $p_k^{(\text{up})}$ and $p_k^{(\text{down})}$, both starting from p_{i_*} and respectively marching towards p_0 and p_{n-1}. Let $\gamma_u > 1$ and $\gamma_d > 1$ denote the ratios of $p_k^{(\text{up})}$ and $p_k^{(\text{down})}$, respectively. Once defined the number of points in each sequence, say n_u and n_d respectively, the following representations are easily obtained:

$$p_k^{(\text{up})} := p_{i_*} + \frac{\gamma_u^{(n_u-1)-k} - 1}{\gamma_u - 1} \, \delta_u \quad , \quad k \in \{0, \ldots, (n_u - 1)\} \qquad \text{(B.1)}$$

$$p_k^{(\text{down})} := p_{i_*} - \frac{\gamma_d^{k-(n_u-1)} - 1}{\gamma_d - 1} \, \delta_d \quad , \quad k \in \{(n_u - 1), \ldots, (n_u + n_d - 2)\} \qquad \text{(B.2)}$$

where:

$$\delta_u := \left(p_0 - p_{i_*}\right) \frac{\gamma_u - 1}{\gamma_r^{(n_u-1)} - 1}$$

$$\delta_d := \left(p_{i_*} - p_{n-1}\right) \frac{\gamma_d - 1}{\gamma_d^{n_d-1} - 1}$$

and the new pressure sequence finally reads:

$$p''_k := \begin{cases} p_0 & , \quad k = 0 \\ p_k^{(up)} & , \quad k \in \{1, \ldots, (n_u - 2)\} \\ p_{i_\star} & , \quad k = (n_u - 1) \\ p_k^{(down)} & , \quad k \in \{n_u, \ldots, (n_u + n_d - 3)\} \\ p_{n-1} & , \quad k = (n_u + n_d - 2). \end{cases} \tag{B.3}$$

The new pressure sequence has a noticeable advantage over the old one: it permits to analytically identify the nodal span to which a given value of the pressure p belongs by inverting (B.1) and (B.2) as follows (the cases $p = p_0$, $p = p_{i_\star}$ and $p = p_{n-1}$ are neglected because trivial):

$$p \in \begin{cases} \left[p''_{\mu(p)}, \, p''_{\mu(p)-1} \right) & , \quad p_{i_\star} < p < p_0 \\ \left(p''_{\nu(p)+1}, \, p''_{\nu(p)} \right] & , \quad p_{n-1} < p < p_{i_\star} \end{cases} \tag{B.4}$$

with:

$$\mu(p) := (n_u - 1) - \left\lfloor \frac{1}{\ln(\gamma_u)} \ln \left\{ 1 + (p - p_{i_\star}) \frac{\gamma_u - 1}{\delta_u} \right\} \right\rfloor \tag{B.5}$$

$$\nu(p) := (n_u - 1) + \left\lfloor \frac{1}{\ln(\gamma_d)} \ln \left\{ 1 + (p_{i_\star} - p) \frac{\gamma_d - 1}{\delta_d} \right\} \right\rfloor \tag{B.6}$$

where, of course, the symbol $\lfloor \cdot \rfloor$ denotes the floor function.

Once defined the new pressure sequence p''_k, a new table can be built either by solving the o.d.e. (4.4) once more, now in correspondence of the sequence p''_k, or by interpolating the original table. The latter strategy is considered here and the following new table, in particular, is built:

$$\left(\rho''_k, \, p''_k, \, a''_k \right) \quad , \quad k \in \{0, \ldots, (n_u + n_d - 2)\} \tag{B.7}$$

by linearly interpolating the original one (4.10) in correspondence of the new pressure sequence (B.3). Clearly, the original table can be discarded at this point, since it is never accessed by the considered algorithm. It may be worth noticing that, besides being attractive for its simplicity, a linear interpolation preserves the strict monotonicity of the p-ρ curve.

For suitable values of the relevant parameters, the new table very well approximates the original one: a fitting practically identical to that one

shown in Figure 4.3 is obtained, not reported here for brevity. It is therefore natural to define the following two-step access strategy based on
table (B.7):

- given an input pressure p (the cases $p = p_0$, $p = p_{i_*}$ and $p = p_{n-1}$
  are not considered here because trivial), the corresponding span within
  (B.7) is identified by means of (B.4)-(B.6);
- the values of ρ and a corresponding to p are defined by linear interpolation within the identified span. Of course, this procedure can be
  extended to an arbitrary number of dependent variables (*e.g.* the function Ψ, defined in (2.64), to be used for solving RPs associated with
  convex state laws, see sec. 2.5.1).

Evidently, the aforementioned access strategy is more efficient than a
crude look-up within the original table (4.10).

References

[1] R. ABGRALL, An extension of Roe's upwind scheme to algebraic equilibrium real gas models. *Computers & Fluids*, 19:171–182, 1991.

[2] R. K. AVVA, A. K. SINGHAL and D. H. GIBSON, An enthalpy based model of cavitation. *ASME-FED*, 226:63–70, 1995.

[3] T. BARBERON and P. HELLUY, Finite volume simulation of cavitating flows. *Computers & Fluids*, 34:832–858, 2005.

[4] A. BERMÚDEZ, A. DERVIEUX, J. A. DÉSIDÉRI and M. E. VÁZQUEZ, Upwind schemes for the two-dimensional shallow water equations with variable depth using unstructured meshes. Rapport de recherche 2738, INRIA, 1995.

[5] F. BEUX, *Conception optimale de formes aérodynamiques et méthodes d'approximations décentrées pour des écoulements incompressibles*. PhD thesis, University of Nice-Sophia-Antipolis, 1993.

[6] F. BEUX, M.V. SALVETTI, A. IGNATYEV, D. LI, C. MERKLE and E. SINIBALDI, A numerical study of non-cavitating and cavitating liquid flow around a hydrofoil. *ESAIM - Mathematical Modelling and Numerical Analysis*, 39(3):577–590, 2005.

[7] M. BILANCERI, Studio dell'effetto della legge di stato nella simulazione di un flusso cavitante barotropico. Tesi di laurea in Ingegneria Aerospaziale, Università di Pisa, Pisa (Italy), a.a. 2005.

[8] P. BIRKEN and A. MEISTER, Stability of preconditioned finite volume schemes at low Mach numbers. *BIT-Numerical Mathemathics*, 45(3):463–480, 2005.

[9] C. E. BRENNEN, *Hydrodynamics of Pumps*. Concepts ETI Inc. and Oxford University Press, 1994.

[10] C. E. BRENNEN, *Cavitation and Bubble Dynamics*. Oxford University Press, 1995.

[11] W. R. BRILEY and H. MCDONALD, An overview and generalization of implicit Navier-Stokes algorithms and approximate factorization. *Computers & Fluids*, 30:807–828, 2001.

[12] H. B. CALLEN, *Thermodynamics and an Introduction to Thermostatistics*. John Wiley & Sons, 1985.

[13] L. CASTELLETTI, G. QUARANTA and L. QUARTAPELLE, Solution of the Riemann problem for van der Waals gas. Technical Report DIA SR 03-04, Dipartimento di Ingegneria Aerospaziale, Politecnico di Milano, Milano (Italy), 2003.

[14] A. CERVONE, L. TORRE, C. BRAMANTI, E. RAPPOSELLI and L. D'AGOSTINO, Experimental characterization of the cavitation instabilities in the AVIO FAST2 inducer. In: *Proc. 41st AIAA/ASME/SAE/ASEE Joint Propulsion Conference*, Tucson (Arizona, USA), July 2005.

[15] Y. CHEN and S. D. HEISTER, Modeling hydrodynamic nonequilibrium in cavitating flows. *Journal of Fluids Engineering*, 118:172–178, 1996.

[16] C.H. CHOI, S.-S. HONG, B.J. CHA and S. YANG, Study on the hydraulic performance of a turbopump inducer. In: *Proc. ASME FEDSM'03 - 4th ASME/JSME Joint Fluids Engineering Conference*, Honolulu (Hawaii, USA), July 2003.

[17] Y.-H. CHOI and C. L. MERKLE, The application of preconditioning in viscous flows. *Journal of Computational Physics*, 105:207–223, 1993.

[18] A. J. CHORIN, A numerical method for solving incompressible viscous flow problems. *Journal of Computational Physics*, 2:12–26, 1967.

[19] A. J. CHORIN and J. E. MARSDEN, *A mathematical introduction to fluid mechanics*. Springer, 1993.

[20] R. COURANT and K. O. FRIEDRICHS, *Supersonic flow and shock waves*. Springer, 1985.

[21] O. COUTIER-DELGOSHA and J.A. ASTOLFI, Numerical prediction of the cavitating flow on a two-dimensional symmetrical hydrofoil with a single fluid model. In: *Proc. CAV2003 - Fifth International Symposium on Cavitation*, Osaka (Japan), November 2003.

[22] O. COUTIER-DELGOSHA, R. FORTES-PATELLA, J. L. REBOUD, N. HAKIMI and C. HIRSCH, Numerical simulation of cavitating flow in 2D and 3D inducer geometries. *International Journal for Numerical Methods in Fluids*, 48:135–167, 2005.

[23] O. COUTIER-DELGOSHA, R. FORTES-PATELLA, J-L. REBOUD, N. HAKIMI and C. HIRSCH, Stability of preconditioned Navier-

Stokes equations associated with a cavitation model. *Computers & Fluids*, 34:319–349, 2005.

[24] O. COUTIER-DELGOSHA, P. MOREL, R. FORTES-PATELLA and J. L. REBOUD Numerical simulation of turbopump inducer cavitating behavior. *International Journal of Rotating Machinery*, 2:135–142, 2005.

[25] O. COUTIER-DELGOSHA, J-L. REBOUD and R. FORTES-PATELLA, Numerical study of the effect of the leading edge shape on cavitation around inducer blade sections. In: *Proc. CAV2001 - Fourth International Symposium on Cavitation*, Pasadena (California, USA), June 2001.

[26] L. D'AGOSTINO, Isenthalpic cavitation model. (Private communication).

[27] L. D'AGOSTINO and E. RAPPOSELLI, A modified bubbly isenthalpic model for numerical simulation of cavitating flows. *AIAA paper 2001-3402*, 2001.

[28] C. DEBIEZ and A. DERVIEUX, Mixed element volume MUSCL methods with weak viscosity for steady and unsteady flow calculation. *Computers & Fluids*, 29:89–118, 2000.

[29] Y. DELANNOY *Modélisation d'écoulements instationnaires et cavitants*. PhD thesis, INPG, Grenoble (France), 1989.

[30] Y. DELANNOY and J. L. KUENY, Cavity flow predictions based on the euler equations. *ASME Cavitation and Multiphase Flow Forum*, 109:153–158, 1990.

[31] A. I. DELIS, C. P. SKEELS and S. C. RYRIE, Implicit high-resolution methods for modelling one-dimensional open channel flow. *Journal of Hydraulic Research*, 5:369–382, 2000.

[32] A. DERVIEUX, Steady Euler simulations using unstructured meshes. Von Karman Institute for Fluid Dynamics, Lecture series 1985-04, 1985.

[33] A. DERVIEUX and J. A. DÉSIDÉRI, Compressible flow solvers using unstructured grids. Rapport de recherche 1732, INRIA, 1992.

[34] L. C. EVANS, *Partial differential equations*. Number 19 in Graduate Studies in Mathematics. American Mathematical Society, 1998.

[35] C. FARHAT, High performance simulation of coupled nonlinear transient aeroelastic problems. Special course on parallel computing in CFD. Technical Report R-807, NATO AGARD, October 1995.

[36] L. FEZOUI and B. STOUFFLET, A class of implicit upwind schemes for Euler simulations with unstructured meshes. *Journal of Computational Physics*, 84:174–206, 1989.

[37] P. GLAISTER, An approximate linearised Riemann solver for the Euler equations for real gases. *Journal of Computational Physics*, 74:382–408, 1988.

[38] P. GLAISTER, A Riemann solver for barotropic flow. *Journal of Computational Physics*, 93:477–480, 1991.

[39] E. GODLEWSKI and P.A. RAVIART, *Numerical approximation of hyperbolic systems of conservation laws*. Number 118 in Applied Mathematical Sciences. Springer, 1996.

[40] S. K. GODUNOV, A finite difference method for the computation of discontinuous solutions of the equations of fluid dynamics. *Matematicheskii Sbornik*, 47:357–393, 1959.

[41] A. GUARDONE and L. VIGEVANO, Roe linearization for the van der Waals gas. *Journal of Computational Physics*, 175:50–78, 2002.

[42] H. GUILLARD and C. VIOZAT, On the behaviour of upwind schemes in the low Mach number limit. *Computers & Fluids*, 28:63–86, 1999.

[43] A. HARTEN and J. M. HYMAN, Self-adjusting grid methods for one-dimensional hyperbolic conservation laws. *Journal of Computational Physics*, 50:235–269, 1983.

[44] C. HIRSCH, *Numerical Computation of Internal and External Flows*, volume 1 (Fundamentals of Numerical Discretizations). Wiley, 1988.

[45] H.W.M. HOEIJMAKERS, M.E. JANSSENS and W. KWAN, Numerical simulation of sheet cavitation. In: *Proc. Third International Symposium on Cavitation*, Grenoble (France), April 1998.

[46] A. HOSANGADI, V. AHUJA and R.J. UNGEWITTER, Simulations of cavitating flows in turbopumps. *AIAA paper 2003-1261*, 2003.

[47] A. HOSANGADI, V. AHUJA and R.J. UNGEWITTER, Simulations of cavitating cryogenic inducers. *AIAA paper 2004-4023*, 2004.

[48] T. Y. HOU and P. LE FLOCH, Why non-conservative schemes converge to the wrong solutions: error analysis. *Mathematics of Computation*, 62(206):497–530, 1994.

[49] U. IBEN, F. WRONA, C.-D. MUNZ and M. BECK, Cavitation in hydraulic tools based on thermodynamic properties of liquid and gas. *Journal of Fluids Engineering*, 124:1011–1016, 2002.

[50] M. ISHII, *Thermo-fluid Dynamic Theory of Two-Phase Flow*. Eyrolles, 1975.

[51] M. J. IVINGS, D. M. CAUSON and E. F. TORO, Riemann solvers for compressible water. In: *Proc. Computational Fluid Dynamics '96 - ECCOMAS*, pages 944–949, 1996.

[52] A. JAMESON, Time-dependent calculations using multigrid, with applications to unsteady flows past airfoils and wings. *AIAA paper 91-1596*, 1991.

[53] A. JAMESON, W. SCHMIDT and E. TURKEL, Numerical solutions of the Euler equations by finite volume methods using Runge-Kutta time-stepping schemes. *AIAA paper 81-1259*, 1981.

[54] M. E. JANSSENS, S. J. HULSHOFF and H. W. M. HOEIJMAKERS, Calculation of unsteady attached cavitation. *AIAA paper 97-1936*, 1997.

[55] A. JEFFREY, *Quasilinear hyperbolic systems and waves*. Pitman, 1976.

[56] A. JEFFREY and T. TANIUTI, *Non-linear wave propagation, with applications to physics and magnetohydrodynamics*. New York Academic Press, 1964.

[57] A. KUBOTA, H. KATO and H. YAMAGUCHI, A new modelling of cavitating flows: a numerical study of unsteady cavitation on a hydrofoil section. *Journal of Fluid Mechanics*, 240:59–96, 1992.

[58] R. F. KUNZ, D. A. BOGER, D. R. STINEBRING, T. S. CHYCZE-WSKI, J. W. LINDAU, H. J. GIBELING, S. VENKATESWARAN and T. R. GOVINDAN, A preconditioned Navier-Stokes method for two-phase flows application to cavitation prediction. *Computers & Fluids*, 29:849–875, 2000.

[59] B. LAKSHMINARAYANA, Fluid dynamics of inducers - a review. *Journal of Fluids Engineering*, 104:411–427, 1982.

[60] P. D. LAX, Development of singularities of solutions of nonlinear hyperbolic partial differential equations. *Journal of Mathematical Physics*, 5:611–613, 1964.

[61] P. D. LAX, *Hyperbolic systems of conservation laws and the mathematical theory of shock waves*. SIAM, 1973.

[62] P. D. LAX and B. WENDROFF, Systems of conservation laws. *Communications On Pure & Applied Mathematics*, 13:217–237, 1960.

[63] R. J. LE VEQUE, *Numerical methods for conservation laws*. Birkhäuser, 1992.

[64] R. J. LE VEQUE, *Finite volume methods for hyperbolic problems*. Cambridge University Press, 2002.

[65] T. LEVI-CIVITA, *Caratteristiche dei sistemi differenziali e propagazione ondosa*. Zanichelli, 1931.

[66] H. W. LIEPMANN and A. ROSHKO, *Elements of Gasdynamics*. John Wiley & Sons, 1957.

[67] R. MARTIN and H. GUILLARD, Second-order defect-correction scheme for unsteady problems. *Computers & Fluids*, 25(1):9–27, 1996.

[68] G. MATTEI, *Lezioni di meccanica razionale*. SEU - Servizio Editoriale Universitario di Pisa, 1995.

[69] R. MENIKOFF and B. J. PLOHR, The Riemann problem for fluid flow of real materials. *Reviews of Modern Physics*, 61(1):75–130, 1989.

[70] C. L. MERKLE, J. FENG and P. E. O. BUELOW, Computational modeling of the dynamics of sheet cavitation. In: *Proc. Third International Symposium on Cavitation*, Grenoble (France), April 1998.

[71] B. N'KONGA and H. GUILLARD, Godunov type method on non-structured meshes for three dimensional moving boundary problems. *Computer Methods in Applied Mechanics and Engineering*, 113:183–204, 1994.

[72] O. OLEINIK, Discontinuous solutions of nonlinear differential equations. *American Mathematical Society translations series*, 2(26):95–172, 1957.

[73] S. A. PANDYA, S. VENKATESWARAN and T. H. PULLIAM, Implementation of preconditioned dual-time procedures in OVER-FLOW. *AIAA paper 2003-0072*, 2003.

[74] S. V. PATANKAR, *Numerical heat transfer and fluid flow*. Hemisphere, 1980.

[75] R. PEYRET and T. TAYLOR, *Computational methods for fluid flows*. Springer, 1983.

[76] B. POUFFARY, R. FORTES-PATELLA and J-L REBOUD, Numerical simulation of cavitating flow around a 2D hydrofoil: a barotropic approach. In: *Proc. CAV2003 - Fifth International Symposium on Cavitation*, Osaka (Japan), November 2003.

[77] A. PRESTON, T. COLONIUS and C. E. BRENNEN, Toward efficient computation of heat and mass transfer effects in the continuum model for bubbly cavitating flows. In: *Proc. CAV2001 - Fourth International Symposium on Cavitation*, Pasadena (California, USA), June 2001.

[78] Q. QIN, C. C. S. SONG and R.E.A. ARNDT, A virtual single-phase natural cavitation model and its application to CAV2003 hydrofoil. In: *Proc. CAV2003 - Fifth International Symposium on Cavitation*, Osaka (Japan), November 2003.

[79] A. QUARTERONI, R. SACCO and F. SALERI, *Matematica numerica*. Springer, 2000.

[80] A. QUARTERONI and A. VALLI, *Numerical approximation of partial differential equations.* Number 23 in SCM Series. Springer, 1994.

[81] E. RAPPOSELLI, A. CERVONE, C. BRAMANTI and L. D'AGO-STINO, Thermal cavitation experiments on a NACA0015 hydrofoil. In: *Proc. ASME FEDSM'03 - 4th ASME/JSME Joint Fluids Engineering Conference*, Honolulu (Hawaii, USA), July 2003.

[82] J-L. REBOUD, B. STUTZ and O. COUTIER, Two-phase flow structure of cavitation: experiment and modelling of unsteady effects. In: *Proc. Third International Symposium on Cavitation*, Grenoble (France), April 1998.

[83] W. C. REYNOLDS, *Thermodynamics properties in SI.* Dept. of Mechanical Engineering, Stanford University, 1979.

[84] P. L. ROE, Approximate Riemann solvers, parameter vectors and difference schemes. *Journal of Computational Physics*, 43:357–372, 1981.

[85] P. ROSTAND, *Sur une méthode de volumes finis en maillage non structuré pour le calcul d'écoulements visqueux compressibles.* PhD thesis, Université da Paris VI, 1989.

[86] Y. SAITO, I. NAKAMORI and T. IKOHAGI, Numerical analysis of unsteady vaporous cavitating flow around a hydrofoil. In: *Proc. CAV2003 - Fifth International Symposium on Cavitation*, Osaka (Japan), November 2003.

[87] I. SENOCAK and W. SHYY, A pressure-based method for turbolent cavitating flow computations. *Journal of Computational Physics*, 176:363–383, 2002.

[88] J. SERRIN, Mathematical principles of classical fluid mechanics. In: *Handbuch der Physik*, volume VIII, pages 125–263. Springer, 1959.

[89] W. SHYY, J. WU and Y. UTTURKAR, Computational modeling of cavitation for liquid rocket applications. *AIAA paper 2004-3985*, 2004.

[90] E. SINIBALDI and F. BEUX, A linearised implicit Roe scheme for inhomogeneous flux functions. In: *Proc. VII Congresso SIMAI - Società Italiana di Matematica Applicata e Industriale (extended abstract)*, Venezia (Italy), September 2004.

[91] E. SINIBALDI, F. BEUX and M.V. SALVETTI, A preconditioned implicit Roe scheme for barotropic flows: towards simulation of cavitation phenomena. Rapport de recherche 4891, INRIA, 2003.

[92] E. SINIBALDI, F. BEUX and M.V. SALVETTI, A preconditioned compressible flow solver for numerical simulation of 3D cavitation

phenomena. In: *Proc. ECCOMAS2004*, Jyväskylä (Finland), July 2004.

[93] E. SINIBALDI, F. BEUX and M.V. SALVETTI, A numerical method for 3D barotropic flows in turbomachinery. *Flow, Turbulence and Combustion*, 76:371–381, 2006.

[94] E. SINIBALDI, F. BEUX, M.V. SALVETTI and L. D'AGOSTINO, Numerical experiments with an homogeneous-flow model for thermal cavitation. In: *Proc. CAV2003 - Fifth International Symposium on Cavitation*, Osaka (Japan), November 2003.

[95] G. A. SOD, A survey of several finite difference methods for systems of nonlinear hyperbolic conservation laws. *Journal of Computational Physics*, 27:1–31, 1978.

[96] C. C. S. SONG AND J. HE, Numerical simulation of cavitating flows by single-phase flow approach. In: *Proc. Third International Symposium on Cavitation*, Grenoble (France), April 1998.

[97] G. P. SUTTON, *Rocket propulsion elements*. John Wiley & Sons, 2001.

[98] E. F. TORO, *Riemann solvers and numerical methods for fluid dynamics*. Springer, 1997.

[99] E. F. TORO, *Shock-capturing methods for free-surface shallow flows*. John Wiley & Sons, 2001.

[100] E. TURKEL, Preconditioned methods for solving the incompressible and low speed compressible equations. *Journal of Computational Physics*, 72:277–298, 1987.

[101] E. TURKEL, Preconditioning techniques in computational fluid dynamics. *Annual Reviews in Fluid Mechanics 1999*, 31:385–416, 1999.

[102] E. TURKEL and V. N. VATSA, Choice of variables and preconditioning for time dependent problems. *AIAA paper 2003-3692*, 2003.

[103] D. R. VAN DER HEUL, *A staggered scheme for nonconvex hyperbolic system of conservation laws*. PhD thesis, Delft University of Technology, 2000.

[104] D. R. VAN DER HEUL, C. VUIK and P. WESSELING, A staggered scheme for hyperbolic conservation laws applied to unsteady sheet cavitation. *Computing and Visualization in Science*, 2:63–68, 1999.

[105] D. R. VAN DER HEUL, C. VUIK and P. WESSELING, Efficient computation of flow with cavitation by compressible pressure correction. In: *Proc. ECCOMAS 2000*, Barcelona (Spain), September 2000.

[106] B. VAN LEER, Towards the ultimate conservative difference scheme III: upstream-centered finite difference schemes for ideal compressible flow. *Journal of Computational Physics*, 23:263–275, 1977.

[107] B. VAN LEER, Towards the ultimate conservative difference scheme IV: a new approach to numerical convection. *Journal of Computational Physics*, 23:276–299, 1977.

[108] B. VAN LEER, Towards the ultimate conservative difference scheme V: a second-order sequel to Godunov's method. *Journal of Computational Physics*, 32:101–136, 1979.

[109] B. VAN LEER, T.-E. LEE and P. ROE, Characteristic time-stepping or local preconditioning of the Euler equations. *AIAA paper 91-1552*, 1991.

[110] S. VENKATESWARAN, J. W. LINDAU, R. F. KUNZ and C. L. MERKLE, Preconditioning algorithms for the computation of multi-phase mixture flows. *AIAA paper 2001-0279*, 2001.

[111] M. VINOKUR and J. L. MONTAGNÉ, Generalized flux-vector splitting and Roe average for an equilibrium real gas. *Journal of Computational Physics*, 89:276–300, 1990.

[112] C. VIOZAT, Implicit upwind schemes for low Mach number compressible flows. Rapport de recherche 3084, INRIA, 1997.

[113] B. WENDROFF, The Riemann problem for materials with nonconvex equations of state, I: Isentropic flow. *Journal of Mathematical Analysis and Applications*, 38:454–466, 1972.

[114] B. WENDROFF, The Riemann problem for materials with nonconvex equations of state, II: General flow. *Journal of Mathematical Analysis and Applications*, 38:640–658, 1972.

[115] G. B. WHITHAM, *Linear and non-linear waves.* John Wiley & Sons, 1974.

[116] J. WU, Y. UTTURKAR and W. SHYY, Assessment of modeling strategies for cavitating flow around a hydrofoil. In: *Proc. CAV2003 - Fifth International Symposium on Cavitation*, Osaka (Japan), November 2003.

[117] T. Y. WU, Cavity and wake flows. *Annual Review of Fluid Mechanics*, 3:243–284, 1972.

[118] H. YAMADA, S. HASEGAWA, M. WATANABE, T. HASHIMOTO, T. KIMURA, J. TAKITA and I. KUBOTA, Observation of the inner flow in the inducer. In: *Proc. 9th International Symposium on Transport Phenomena and Dynamics of Rotating Machinery*, Honolulu (Hawaii, USA), February 2002.

THESES

This series gathers a selection of outstanding Ph.D. theses defended at the Scuola Normale Superiore since 1992.

Published volumes

1. F. COSTANTINO, *Shadows and branched shadows of 3 and 4-manifolds*, 2005. ISBN 88-7642-154-8

2. S. FRANCAVIGLIA, *Hyperbolicity equations for cusped 3-manifolds and volume-rigidity of representations*, 2005. ISBN 88-7642-167-x

3. E. SINIBALDI, *Implicit preconditioned numerical schemes for the simulation of three-dimensional barotropic flows*, 2007.
ISBN 978-88-7642-310-9

Volumes published earlier

H.Y. FUJITA, *Equations de Navier-Stokes stochastiques non homogènes et applications*, 1992.

G. GAMBERINI, *The minimal supersymmetric standard model and its phenomenological implications*, 1993. ISBN 978-88-7642-274-4

C. DE FABRITIIS, *Actions of Holomorphic Maps on Spaces of Holomorphic Functions*, 1994. ISBN 978-88-7642-275-1

C. PETRONIO, *Standard Spines and 3-Manifolds*, 1995.
ISBN 978-88-7642-256-0

I. DAMIANI, *Untwisted Affine Quantum Algebras: the Highest Coefficient of* det H_η *and the Center at Odd Roots of 1*, 1996.
ISBN 978-88-7642-285-0

M. MANETTI, *Degenerations of Algebraic Surfaces and Applications to Moduli Problems*, 1996. ISBN 978-88-7642-277-5

F. CEI, *Search for Neutrinos from Stellar Gravitational Collapse with the MACRO Experiment at Gran Sasso*, 1996. ISBN 978-88-7642-284-3

A. SHLAPUNOV, *Green's Integrals and Their Applications to Elliptic Systems*, 1996. ISBN 978-88-7642-270-6

R. TAURASO, *Periodic Points for Expanding Maps and for Their Extensions*, 1996. ISBN 978-88-7642-271-3

Y. BOZZI, *A study on the activity-dependent expression of neurotrophic factors in the rat visual system*, 1997. ISBN 978-88-7642-272-0

M.L. CHIOFALO, *Screening effects in bipolaron theory and high-temperature superconductivity*, 1997. ISBN 978-88-7642-279-9

D.M. CARLUCCI, *On Spin Glass Theory Beyond Mean Field*, 1998. ISBN 978-88-7642-276-8

G. LENZI, *The MU-calculus and the Hierarchy Problem*, 1998. ISBN 978-88-7642-283-6

R. SCOGNAMILLO, *Principal G-bundles and abelian varieties: the Hitchin system*, 1998. ISBN 978-88-7642-281-2

G. ASCOLI, *Biochemical and spectroscopic characterization of CP20, a protein involved in synaptic plasticity mechanism*, 1998. ISBN 978-88-7642-273-7

F. PISTOLESI, *Evolution from BCS Superconductivity to Bose-Einstein Condensation and Infrared Behavior of the Bosonic Limit*, 1998. ISBN 978-88-7642-282-9

L. PILO, *Chern-Simons Field Theory and Invariants of 3-Manifolds*, 1999. ISBN 978-88-7642-278-2

P. ASCHIERI, *On the Geometry of Inhomogeneous Quantum Groups*, 1999. ISBN 978-88-7642-261-4

S. CONTI, *Ground state properties and excitation spectrum of correlated electron systems*, 1999. ISBN 978-88-7642-269-0

G. GAIFFI, *De Concini-Procesi models of arrangements and symmetric group actions*, 1999. ISBN 978-88-7642-289-8

N. DONATO, *Search for neutrino oscillations in a long baseline experiment at the Chooz nuclear reactors*, 1999. ISBN 978-88-7642-288-1

R. CHIRIVÌ, *LS algebras and Schubert varieties*, 2003. ISBN 978-88-7642-287-4

V. MAGNANI, *Elements of Geometric Measure Theory on Sub-Riemannian Groups*, 2003. ISBN 88-7642-152-1

F.M. ROSSI, *A Study on Nerve Growth Factor (NGF) Receptor Expression in the Rat Visual Cortex: Possible Sites and Mechanisms of NGF Action in Cortical Plasticity*, 2004. ISBN 978-88-7642-280-5

G. PINTACUDA, *NMR and NIR-CD of Lanthanide Complexes*, 2004. ISBN 88-7642-143-2

Fotocomposizione "CompoMat" Loc. Braccone, 02040 Configni (RI) Italy
Finito di stampare nel mese di luglio 2007 presso
BRAILLE-GAMMA s.r.l. Cittaducale (RI)